Landforms of High Mountains

Alexander Stahr • Ewald Langenscheidt

Landforms of High Mountains

Springer

Alexander Stahr
Taunusstein
Germany

Ewald Langenscheidt
Rotthalmünster
Germany

ISBN 978-3-642-53714-1 ISBN 978-3-642-53715-8 (eBook)
DOI 10.1007/978-3-642-53715-8
Springer Heidelberg New York Dordrecht London

Library of Congress Control Number: 2014945812

Peitlerkofel (Sass de Putia, 2875 m). Highest mountain of Peitlerkofelgruppe in the Dolomites (Italy).
On June 26, 2009, parts of the Dolomites were recognized by UNESCO (United Nations Educational, Scientific and Cultural Organization) as a serial World Heritage Site. The Peitlerkofel massif is a UNESCO World Heritage and includes all typical rock formations of the Dolomites.

Printed on acid-free paper

Springer is part of Springer Science+Business Media (www.springer.com)

Preface

When the high mountains call, then they come: The hikers and mountaineers, pupils, students, scientists, the juniors, the seniors, and the moderate or the extreme ones. All of them come and they come with pleasure. There are no exact numbers, but it might be millions annually who travel into the high mountains of the world.

All of us enjoy the beauties of the mountains: We take photographs, walk their valleys, climb over their glaciers, and climb their walls. We think often we know what we see at different landforms. However, can we also describe or explain them? And what actually do we know about their emergence? But what are the things called which we see? What exactly can we say about the various landforms? To us names like Reinhold Messner, Sir Edmund Hillary, Ueli Steck, Gerlinde Kaltenbrunner, or Wanda Rutkiewicz are familiar. But do we know the terms "lapies" or "shore moraine"?

This book shall give information about the various landforms in the high mountains. It is a textbook which is understandable and illustrated richly, an adviser, as well as a reference book. The book therefore addresses both the interested layman and pupils and teachers or students of earth sciences for introduction to the geomorphological variety of high mountain areas. However, it cannot and will not replace any conventional textbook. The secondary documents are referred to in the bibliography to deepen studies of the topics dealt with.

Most given examples are from Europe, Asia, and the North and South Americas as well as from Oceania. Completeness was not striven for. Anyway it would be impossible to discuss comprehensively the enormous variety of landforms in the high mountains in one book. Photos and textual content therefore confine themselves to a subjective choice.

Our thanks go to Herbert Funk (Frankfurt am Main, Germany), Heiko Kurz (Mainz, Germany), Dr. Johannes Thomas Weidinger (Gmunden, Austria), and Prof. Dr. Heinz Veit (University of Bern, Switzerland) for leaving photographic materials. The excellent photos of Herbert Funk from Svalbard, Norway, and Greenland can be seen on the Web site http://www.arcticimpressions.de. A wealth of photographs and other materials can be found at the U.S. Geological Survey (http://www.usgs.gov) that can provide further insights into the world of high mountains' geomorphology.

With this book we hope to give to a broader public a better understanding of the fascinating form variety in the high mountains and also an appreciation of the essential causal geomorphological relationships in the high mountains region.

Taunusstein, Germany
Rotthalmünster, Germany
Spring 2014

Alexander Stahr
Ewald Langenscheidt

Contents

Introduction

1

A. Stahr and E. Langenscheidt, *Landforms of High Mountains*,
DOI 10.1007/978-3-642-53715-8_1, © Springer-Verlag Berlin Heidelberg 2015

They can be found on all continents; their beauty is raw and alluring, but equally forbidding and threatening. They extend over thousands of kilometers in one direction but only hundreds or tens of kilometers in the other: the high mountain regions of our planet. The word "mountain" triggers a variety of associations: one might think of a particular mountain or a region in which they have been, or have always dreamed of visiting — the Alps, Rocky Mountains, Himalaya, or the mountains of Scandinavia. Legendary mountains such as the Matterhorn, Mont Blanc, Mount McKinley, the 8,611 m high K2, or perhaps the Drei Zinnen (Three Peaks) in the Dolomites come to mind. One thinks perhaps of the weather-beaten alpine farmer, who must fight adverse conditions in order to barely secure his existence. Or perhaps one might think of the exotic people who inhabit earth's highest regions: the Berbers, Balti, Hunzas, Tibetans, and others, whose lives are characterized by a sparse seed in barren earth. Perhaps it is the vision of the fearless mountaineer conquering vertical faces and jagged peaks and venturing into the hostile realm that joins earth with the heavens that fuels one's mountain fantasy. The most fundamental associations with the word mountain are surely the inexhaustible and rugged forms of high elevations and the forces of nature affiliated with them: white-water streams or massive glaciers grinding their way slowly to the valley floor.

1.1 High Mountains and Man

In the minds of most, mountains represent a landscape rich in natural beauty — a refuge of solitude and in some cases even a natural playground. However, the cultural significance of high mountains differs among the various societies of mankind (Fig. 1.1). During the pre-Columbian period, the Incas made sacrificial offerings on the summit of the 6,739 m high Llullallaco in the Chilean Andes. During the same period in Europe, the Alps were a no-man's-land avoided by those god-fearing people. Demons and ghosts allegedly inhabited this ungodly landscape, the Church even condemning it as the forecourt of hell. In the past and present, the peaks of the Himalayas are the most sacred place for the Sherpas and other mountain tribes. Worshiped as the "throne of the gods," some peaks, such as the Machapuchare in the Annapurna area, are so holy that no mortal foot should be set upon them. Mount Fuji, on the other hand, is a site of pilgrimage for Japanese Shintos, as is the Mount Ruapehu volcano for the Maori in New Zealand. The Greeks believed Mount Olympus to be the home of the gods, and the Bible describes Noah's ark landing at the 5,165 m high volcano Mount Ararat in the highlands of Turkey.

What reasons make mountains so important to mankind though? Why are they the "throne of the gods," why did their summits serve as sacrificial altars, why is almost every Alpine peak crowned with a Cross (Fig. 1.2), and why do devout Buddhists and Hindus pilgrim to the holy Mount Kailash, encircling it and spending two weeks crawling around the base on their hands and knees? Without a doubt, the earliest interactions between mountain and man were of religious and mythical nature. It is somewhat a paradox that on the one hand the highest mountain regions were avoided anxiously by those god-fearing, but on the other hand it was assumed that ascension to "gods' throne" itself was proof of devotion. Of the endless landscapes on Earth, mountains are with their jagged peaks, steep faces, massive glaciers, and deep gorges, the most impressive. Mountains left such a lasting impression on mankind, leaving people of all religions and beliefs in a state of awe that they ultimately became the core of human myth and cult. There are many examples of mystical or religious interpretations of numerous seemingly inexplicable mysterious landscapes and natural phenomena. Potholes, for example, were interpreted by alpine people as "devil's mills" or "witches' cauldrons." Glacial erratics were called druid, witch, or devil's stones and the movement of glaciers was attributed to mysterious forces of magic. The Swiss Bernina group is home to the Lej Sgrischus, "the horrible lake." A peculiarly jagged peak in the Ampezzo Dolomites is referred to as "Witch Rock." In the languages of Tibet and Nepal, Mount Everest is called Chomolungma Sagarmatha, meaning "mother of the world".

Since the early human colonization of the mountains — of which the oldest records in the Alps date back to 100,000 years before present — man was obliged to adapt to the forces of nature; regardless of whether these forces left believers terror-stricken or awestruck, they presented an immutable part of the natural habitat. For the advanced cultures of the ancient world, mountains were nothing more than obstacles to be overcome. The Romans regarded the mountains with unequivocal disdain; the term "montes horribiles" coined by Roman literates speaks for itself. This sentiment developed further in the Middle Ages into a deeply instilled fear of mountainous regions. Eventually, mountains were regarded as nothing less than the forecourt to hell. For this reason, accounts of alpine travels during medieval times contain no testimony regarding the natural beauty and splendor of the diverse landscapes. For centuries, the lack of scientific knowledge about the high mountains left man ignorant and vulnerable to nature and its elements. Progress in this respect was hampered by the influence of the Church until the nineteenth century (Fig. 1.3).

Towards the end of the eighteenth century, the Scottish naturalist and geologist James Hutton (1726–1797) visited the Swiss Alps. He found boulders not coherent with the lithology of the surrounding mountains. After detailed landscape observations, Hutton concluded that glaciers upon which he saw many such boulders must flow down the mountain transporting their freight to the valley floor. As

Fig. 1.1 Holy Mountain. The picture shows the 2,518 m high Mount Egmont (or Taranaki) on the west coast of New Zealand's North Island. According to Maori mythology, the volcano is a sacred mountain god. With an age of 10,000 years, it is the youngest member of New Zealand's volcanic arc. The distinct conical shape of this typical stratovolcano has led many Japanese tourists to refer to it as "New Zealand's Mount Fuji." It even served as the backdrop in director Edward Zwick's 2003 film "The Last Samurai." Published with kind permission of © Dr. Ewald Langenscheidt 2014. All Rights Reserved

Fig. 1.2 Holy Cross. Rifugio Lorenzi (2,940 m), Cristallo group, Dolomites, Italy. Why are so many alpine peaks festooned with a holy Cross? The inscription found on so many alpine altars sheds light upon the matter: "Many roads lead to God, one of them over the mountains." This is likely a demonstration of the superiority of religious power over the high mountains, which for centuries were perceived as hell on earth by god-fearing Europeans. Published with kind permission of © Dr. Alexander Stahr 2014. All Rights Reserved

the glacier broke up, proposed Hutton, the boulders were deposited far from their origins. This view was not widely accepted at this time, because the erratic boulders were considered by the Church as irrefutable evidence for the Noachian deluge, which had rolled the boulders over hills and valleys. Gradually, however, a number of contemplative

Fig. 1.3 Cliff. Mount Peitlerkofel (2,875 m), Dolomites, Italy. Vertical rock walls hundreds of meters high fascinate and allure. Landscapes such as these are threatening for many. Even a member of modern civilization might very well feel tense and intimidated in the face of such overwhelming structures. What must ancient people have felt when confronted with such imposing formations? Published with kind permission of © Dr. Alexander Stahr 2014. All Rights Reserved

open-minded people including scientists, hunters, bridge builders, and loggers came to the same conclusion as Hutton. Many observers noticed the strangely flat, almost polished bottom sides of the boulders. In their eyes, it seemed to be more logical that this could be the work of glaciers and not that of biblical cataclysms. However, even these wise observers were not yet influential enough to dispel the believers from biblical explanations for landscape forms (Fig. 1.4).

At a meeting of the Swiss Society of Natural Sciences in Neuchâtel in June 1837, the brilliant Swiss-American zoologist, paleontologist, and geologist Jean Louis Rodolphe Agassiz (1807–1873) held the opening speech. Addressing an assembly of Europe's leading natural scientists, Agassiz proposed that no flood had moved the boulders but rather glaciers. Furthermore he postulated that the valley in which they can be found was once buried under a one-and-a-half kilometer thick layer of ice. He further concluded that glaciers must have once covered all of Europe. The attending scientists and members of the society were outraged by Agassiz's postulations. He requested that the conference attendees follow him into the field, in order to examine the evidence for his case in situ. Agassiz was able to convince some that very day, but that was merely the beginning of a strenuous international campaign to convince the last skeptics that glaciers flow and move and that they had spread throughout much of Europe, causing a wide range of phenomena in mountain landscapes.

Perhaps it was also mountaineering which played an influential role in the demystification of high mountains and sparked interest in their forms and elements. The year 1336 marks the birth of modern alpinism. The lyric poet Francesco Petrarca ascended "animated alone by the desire to learn about the extra height" with his younger brother Girardo on the April 26, the 1,912 m high Mont Ventoux, a foothill of the Alps (Dauphiné Alps) in the Provence, France. Hunters, shepherds, and farmers likely traveled over higher mountains long before 1336, but as Petrarca's account was the first written, history has honored him as the "father of alpinism" (Fig. 1.5).

Over the next few centuries more and more people shed their fear of the high mountains and allowed another human trait to take control: curiosity. In the late fifteenth century, none other than Leonardo da Vinci (1452–1519) undertook extensive hikes in the Monte Rosa area of Italy. Da Vinci's motivation was not only the measurement of elevation or the observation of optical phenomena, but also to simply enjoy the natural scenery which is indisputably reflected in his paintings. In 1511 Leonardo climbed the 2,556 m high Monte Bo of the Monte Rosa Massif. The desire for scientific progress and the fascination for alpinism merges even stronger, as illustrated by the Swiss naturalist Conrad Gessner (1516–1565) in a letter to friends: "As long as god gives me life, I shall be determined to climb some mountains — or at least one mountain every year — and was when the mountain flora blooms fullest — not only to investigate, but also

Fig. 1.4 Erratic Boulder. Northern Limestone Alps, Berchtesgadener Land, Germany. The occasional occurrence of large boulders in the middle of a wide valley floor riddled farmers for ages. They attributed the existence of these boulders to dark forces and sorcery. Names such as "druid stone," "witch stone," or "devil's stone" speak for themselves. Published with kind permission of

Fig. 1.5 Mont Ventoux, France. The Petrarca brothers' ascent of Mont Ventoux in 1336 goes down in history as the first instance of Alpine romantic; the natural experience in the high mountains became an end in itself. Published with kind permission of

Fig. 1.6 Glacier. Briksdals Glacier, Norway. Glaciers have riddled humans for thousands of years. Only the bravest souls dared to set foot on a glacier. For the well-equipped, experienced climbers of modern times crossing a glacier is common play. Even after it was discovered that glaciers move, it remained a mystery as to why they move. Mysterious or magical powers were common explanations. The Swiss scientist Horace Bénédict de Saussure (1740–1799), one of the first to climb Mont Blanc, was also one of the first scientists who presented evidence that glaciers are indeed mobile, describing the influence of gravity on the flow of ice. Published with kind permission of

to provide my body with the noblest of exercises and my spirit with joy. What delight, what bliss the soul experiences when admiring the prodigious massifs and extending one's pate into the clouds! The feeble-minded cower sedentarily in their dwellings, writhing in dung, confused by profit and desire; the disciples of wisdom however, shall always feast their mind's eye upon the drama of earthly paradise ...".

Soon thereafter not only mountaineering but also scientific and cultural development in the German-speaking world suffered a hindering setback. Social, religious, and dynastic wars derailed progress that had become so promising. When peace finally returned to central Europe, interest in the inhospitable Alps or other mountains had subsided. The European eye became beguiled by the romantic gardens, the opulent avenues of Versailles, where nature became a whimsical decoration, a stage set for the wealthy. However, nature itself was at least partially responsible for the blind eye which Europe turned on the Alps. During the little ice age of the sixteenth to seventeenth century, glaciers blocked many alpine passes formerly traversable. It was not until the eighteenth century that mountaineering and the natural experience alone began to regain popularity. The Alps transformed from the ungodly "monte teribillis" into beautiful mountains and hence alpine romanticism was born. The industrialized society designed an idyllic contrast to itself which in essence still exists today, despite the environmental, economic, and political problems of earth's mountainous regions (Fig. 1.6).

In 1802, the omnipresent man of science Friedrich Heinrich Alexander Freiherr von Humboldt (1769–1859) attempted to climb the 6,310 m high Chimborazo in Ecuador. Although he failed in his attempt, Humboldt reached an altitude of 5,893 m, which remained a human elevation record long thereafter. However, Humboldt's motivation was much more than to simply conquer the mountain, but more importantly to study the mountain's vulcanologic properties. He also presented comparative studies in the high mountains concerning summit heights, mountain passes, plateaus, waterfalls, and the limits of perpetual snow. It is the swiss scientist Horace Bénédict de Saussure (1740–1799), however, who is regarded as the founder of alpine research. De Saussure was one of the first on the Mont Blanc, as well as one of the first to provide a plausible explanation for the movement of glaciers. It was de Saussure who recognized the influence of gravity on the movement of large bodies of ice, dispelling the mysterious or magical powers once thought to be responsible.

A large number of scientists, including the swiss geologist Albert Heim (1849–1937), the German geographer Eduard Brückner (1862–1927), and Albrecht Penck (1858–1945), brought forth yet more fundamental insight regarding the characteristics, elements, and shapes of the mountains. It must be noted however that many landscape forms found in the high mountains are of anthropogenic origin. For example, flattened slopes, which reduce soil

Fig. 1.7 Cultural landscape. Großglockner High Alpine Road, Austria. Only the summit regions and the "eternal ice" can be considered, if only partially natural landscapes. Even those spots untouched by agriculture and inaccessible to mountaineers are still victims to air pollution. In addition, overfertilization caused by increased atmospheric nitrogen concentrations has altered the composition of species in plant communities. Published with kind permission of

erosion, are present in most mountainous regions of the world. In some regions of the Himalaya and the Western or Southern Alps, such terrace landscapes are definitive (Fig. 1.7).

As a result of extensive clear-cutting in order to win pastureland in subalpine zones, the timberline was reduced about 200–400 m in height in the Alps. Despite intensive care of pastures in the centuries past, the various forms of erosion that occur in alpine regions can be described as seminatural at most. This means that erosional processes once initiated by man, due to the agricultural usage of areas formerly forested, eventually return to a more or less naturally regulated state. Even large-scale landslides and the characteristic landscapes they create are often due to anthropogenic influence. The Alps bear countless examples: the landslide near Elm of 1881 in the Swiss canton of Glarus or the rockslide that overtopped the Italian Vaiont Dam in 1963. Extensive slate mining caused the catastrophe near Elm, and careless engineering during the construction of the Vaiont Dam caused devastating consequences in the valley.

Man has also transformed the valleys of many high mountain landscapes dramatically. Floodplains were drained in order to be cultivated, rivers were straightened, and streams dammed or diverted. The never-ending demand for building materials and natural resources "gnawed" away entire mountains. Abandoned quarries several hundred meters deep are the legacy that remains. When hiking or climbing in developed mountain regions, only the fewest are aware that they are not taking pleasure in pristine wilderness but in a cultural landscape. A prime example of a high mountain cultural landscape are the Alps; in terms of infrastructure, there is no mountain region better developed on earth (Fig. 1.8).

Although science has explained virtually all natural phenomena of the high mountains and even though we possess a much deeper insight into the complex processes and interactions of formation than our ancestors did, the allure of mountains remains irresistible to man. The extraordinarily wide spectrum of high mountain landscapes and the spatially confined abundance of fascinating and bizarre landforms annually attracts millions of people to the elevated regions of the world.

In many places, however, tourism goes hand in hand with the environmental, economic, and cultural problems it ultimately brings. In contrast to our ancestors, for modern man the mountains represent an omnipresent significance: as the last wilderness, as a recreational area, as an economic factor for energy and natural resources, as a national identity, or as home and habitat. In great contrast to "classical alpinism" for many, the high mountains have become merely a playground. They have been degraded to mountain bike, paraglider, snowboard, or rafting park. Even worse, if the landscape should not fulfill the developer's vision, it shall be reshaped until it does. Ski slopes are flattened, vertical faces are crowned with observation decks, trails are lined with handrails, and hotels and spas are built.

Fig. 1.8 Torrent control. Danger posed by mountain streams is controlled by obstruction, which often causes a distinct landscape change. Published with kind permission of

1.2 What Does "High Mountain" Mean?

This book deals with the high mountains and their landscapes, but what is a high mountain? How high is high? The question may seem a bit trite at first, but it is in fact entirely justified. The central Tibetan plateau with a mean elevation of about 5,000 m or the 4,000 m high Altiplano of Bolivia lack many high mountain characteristics despite their great altitude. They are in fact nothing more than (very) high plains. In contrast the jagged Alpinesque peaks of Svalbard or the rugged Lofoten islands rise directly out of the ocean and are no higher than 2,000 m. Using sea level as a frame of reference, in terms of altitude the peaks of Svalbard or the Lofoten are better compared with the low mountain peaks of central Europe, such as the 1,493 m Feldberg in the Black Forest in Germany.

Intense relief with steep walls and sharp ridges is not the only criteria for a high mountain region. High cliffs above deep river valleys can be found elsewhere, for example, in the Grand Canyon of Arizona. The same applies for glaciation; if applied as a high mountain criterion, then the French Vosges and the Black Forest would be falsely classified as high mountains. The characteristic kar shapes of these mountains are merely the result of glaciation during the last ice age. Glacier bearing volcanoes, such as the 5,895 m Mount Kilimanjaro in East Africa, are considered to be high mountains. Hawaii's Mauna Loa on the other hand is not glaciated and should therefore not be classified as a high mountain if glaciation is the decisive criterion, despite a height of at least 4,170 m above sea level. Furthermore, Mauna Loa shows altitudinal zonation, which is decisive in defining high mountain. All high mountains, even those in Greenland or the Antarctic, display a distinct vertical organization of environments (Fig. 1.9).

To discuss the many high mountain definitions, formulated in detail by geographers, geoscientists, or ecologists, would exceed the scope of this book. The concept of altitudinal belts, however, is a reoccurring aspect in the majority of these definitions. Young uplift, high relief energy, rapid exhumation, high tectonic stress, and extension are justified amendments to the altitudinal belt criteria. In this right, not only mountain ranges and massifs can be defined as high mountains, but also high lone peaks such as the volcanoes Mauna Loa or Kilimanjaro. In summary, high mountains are parts of the earth's crust, which are characterized by high relief energy and in particular by spatially confined changes in environment with altitude, i.e., a clear climate-related stair-step progression of environment.

1.3 The Origin of High Mountains

Many landforms found in the high mountains are directly related to the tectonic processes that formed the mountains. Therefore, an understanding of orogenesis (the formation of mountains) is instrumental in understanding the wide variety

Fig. 1.9 High mountains. Allgäu, Germany. A vertical succession of altitudinal zones is characteristic of high mountain regions. The timberline is clearly visible in the picture. Published with kind permission of © Dr. Alexander Stahr 2014. All Rights Reserved

of morphological manifestations it creates. In order to understand orogenesis, one must know that the earth is not a rigid body. It is composed of several concentric shells that vary greatly according to temperature, aggregate state, chemical composition, and density. The Earth can be divided roughly into crust, mantle, and core. The existence of such shells has been confirmed by seismological techniques. The technique relies upon the analysis of seismic waves (from the Greek word seismos = concussion), whose velocity as well as reflective or refractive properties is dictated by the media through which they travel.

The Earth's exterior shell consists of continental and oceanic crust. The continental crust is 25–50 km thick and is build up of a relatively thin layer of sedimentary rocks and relatively acidic rocks of volcanic or magmatic origin with a specific gravity of 2.7 g/cm^3. Due to their high silicon and aluminum contents, such rocks are described as "sial." The oceanic crust varies only slightly in thickness between 5 and 8 km. It is composed mainly of dense basaltic rocks with a specific gravity of around 3.0 g/cm^3. Silicon and magnesium compounds prevail here ("sima") in addition to a thin sedimentary cover. Earth's crust and uppermost part of the mantle form the lithosphere (from the Greek words lithos = stone and sphaira = ball). As the outermost shell of the earth, the 70–120 km thick lithosphere is relatively cool, causing it to behave brittle. The next shell, the asthenosphere (from the Greek word asthenos = soft), behaves plastically under the high confining pressure and temperatures of 1,000–1,200 °C. The lower part of the upper mantle is the mesosphere (from the Greek mesos = middle). It lies at a depth of about 350–700 km. To a depth of about 2,900 km follows the lower mantle, and finally the outer and inner core. The outer core is liquid and extends to a depth of 5,000 km. Despite temperatures up to 6,000 °C, the inner core is solid and consists of iron–nickel alloy.

The lithosphere is divided into roughly one dozen plates. The boundaries of the plates are not always identical with continental boundaries. Some plates consist of continental as well as oceanic crust, whereas others are only oceanic. The plates "float" on the denser asthenosphere and drift at speeds of up to 10 cm per year (Fig. 1.10).

Because the plates move independently of each other they either collide at their boundaries or drift apart. Plates moving away from each other are separated by divergent boundaries. Along the resulting rift, mantle material ascends and is subject to decompression melting in the process. The result is a magma reservoir below the rift system out of which the magma ascends penetrating the cold overlying lithosphere. In this manner new crust is born, the diverging plates growing centimeter by centimeter. Depending on where the plate boundaries run, this can happen under oceanic or continental crust.

On the ocean floor divergent boundaries are manifested as ridges. "Seafloor spreading" describes the formation of new oceanic crust at the ridges. Due to the low density of the ascending magma, the seabed swells up to a huge ridge with a central trench or rift valley. As the hot material moves away from the ridge, it cools and becomes denser. In the course of cooling the crust fractures into large blocks, which slip down in relation to the adjacent block as they move away from the ridge. At some places, the ridges are so high that they create volcanic islands. Iceland, with Europe's

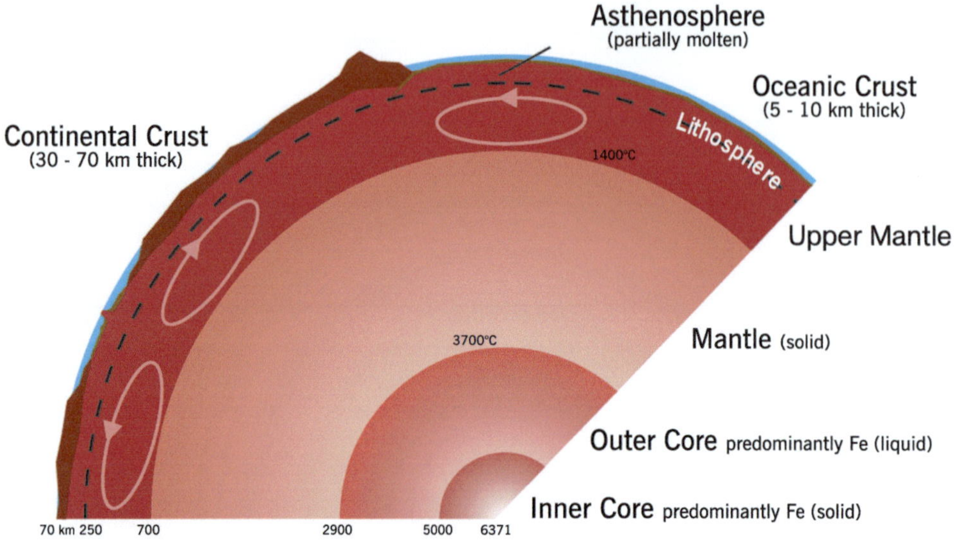

Fig. 1.10 Shell structure of the earth (schematically). The Earth is made up of several concentric shells that vary significantly in temperature, aggregate state, chemical composition, and density from each other. Published with kind permission of © Dr. Ewald Langenscheidt 2014. All Rights Reserved

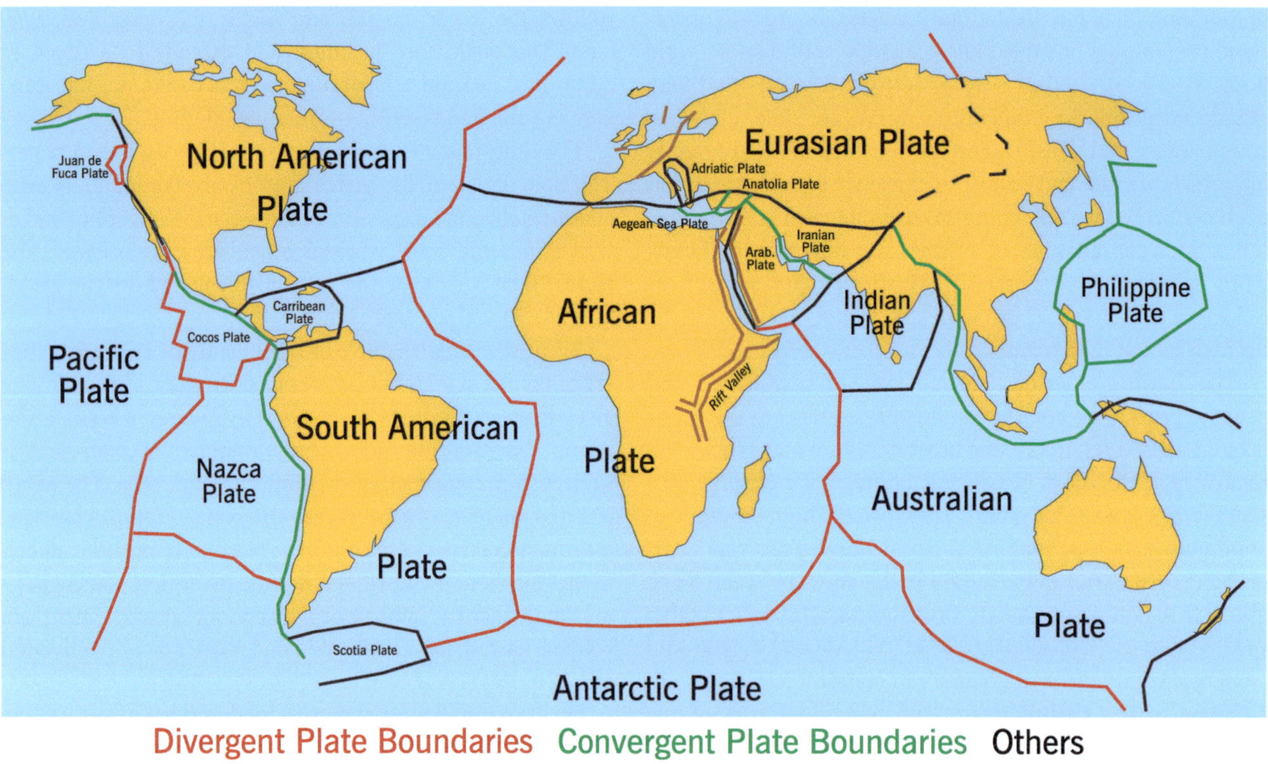

Fig. 1.11 Lithosphere plates. The cool outer shell of the Earth is broken into many plates. Published with kind permission of © Dr. Ewald Langenscheidt 2014. All Rights Reserved

largest glacier, the 8,456 km^2 Vatnajökull, is part of the mid-Atlantic ridge, which extends north to south across the entire length of the Atlantic (Fig. 1.11).

The breakup of a continent is also characterized by extensive rift valleys with volcanism and graben tectonics. Volcanoes with high-alpine characteristics such as Mount Kilimanjaro (5,895 m) or Mount Kenya (5,199 m) line the African Rift Valley.

The divergence of lithospheric plates is offset by plate collision. In the case of collision, one must distinguish

Fig. 1.12 Mount Everest. The 8,848 m high Mount Everest (Sagarmatha in Nepali language, Tibetan: Chomolungma) is the result of the collision between the Indian and Eurasian plates. Published with kind permission of © Wikipedia (public domain) 2014. All Rights Reserved

between oceanic and continental plate margins. The heavier oceanic plates always go under the lighter continental plates; this process is called a subduction (from the Latin subducere = pull down). Subduction results in deep-sea trenches which may have depths of several kilometers. Examples are the 8,066 m deep Atacama trench in front of the west coast of South American or the 6,662 m deep Central American trench in front of the Pacific coast of Mexico and Guatemala. The edge of the continental plate is compressed during collision, folded, and uplifted. The plate being subducted into the mantle melts at depths of about 100 km. Magma rises, creating a coastal volcanic range, which is characteristic for subduction zones. In addition, liquid and gas phases of the water-saturated ocean floor sediments make the melt more buoyant. The great volcanic peaks of the Andes, such as Chimborazo (6,310 m) and Cotopaxi (6,005 m), are typical examples of subduction volcanoes.

When two oceanic plates collide, one must give way to the other and become subducted. Magma rises as the subducted plate melts and volcanic island arc along the plate boundary results. The high volcanoes of the Philippines are a prime example of this phenomenon. When two continental plates meet, however, neither will give way because of their low density. In this case, one plate is pushed over the other resulting in increased crustal thickness and leading to folding and nappe tectonics. The Alps for example were created by such a continent–continent collision and are hence the namesake for this particular style of orogenesis (from the Greek word soros = mountain and genesis = derived). Alpline-style orogenic structures include the Himalaya with the 8,848 m high Mount Everest, the highest mountain in the world. The Himalayas were created during the collision of India with the Eurasian plate (Fig. 1.12). The formation of high mountains is thus based upon the opening and closing of oceans, the formation of new oceanic crust at mid-oceanic ridges, and the subduction of older oceanic crust in subduction zones. High mountains created by hot-spot activity, such as the 4,202 m high Mauna Kea in Hawaii, are the exception.

Orogenesis with the exception of hot spots can be broken down into seven developmental stages (A–G), which is referred to as the Wilson cycle, named after the Canadian geophysicist and geologist John Tuzo Wilson (1908–1993):

(A) A tectonically stable continent exists, also called a craton. It is bordered by ocean basins and eroded more or less to a peneplain. With little or no tectonic activity, this stage can last for tens to hundreds of million years.
(B) Orogenesis is preceded by a stage in which the continent rifts into different plates by intrusive rocks from the mantle. Extension favors the formation of fault systems where magma from the mantle can easily penetrate the crust. As volcanic activity and faulting increase, large rift valleys are created as is the case today in east Africa.
(C) As extension endures, the rift valleys become so deep and extensive that sooner or later the ocean intrudes, forming a basin separating the continental fragments from the craton. The magmatic activity that once took place on the rift valley floor now forms a mid-oceanic ridge. The term "seafloor spreading" describes the genesis of oceanic crust at such divergent boundaries. The Red Sea is a prime example of such an infantile oceanic basin.
(D) As seafloor spreading continues, the oceanic basin grows and the continents drift further apart. The result

is a wide oceanic basin bordered by continents with passive margins. Hot young oceanic crust cools as it moves away from the mid-oceanic ridge. As cooling progresses, the density and thickness of the oceanic crust increase. Hot, young crust is buoyant with respect to the mantle on which it floats, whereas the burden of the colder, thicker, old crust leads to subsidence at the continental margins where sedimentation occurs. The Atlantic Ocean with its mid-oceanic ridge system is an example of a mature oceanic basin.

(E) The oceanic crust at the margins of a mature ocean basin becomes so heavy that forces of subsidence overrule those of buoyancy. The result is the so-called spontaneous subduction, where the brittle oceanic crust shears off in front of the once passive continental margin. Because oceanic crust has a greater density than continental crust, it becomes subducted. Whereas new crust is created at divergent boundaries, old crust is destroyed at such convergent boundaries. Roughly speaking, subduction is powered by two forces: seafloor spreading pushes the oceanic plate from the mid-oceanic ridge, i.e., "ridge push," and the weight of the already subducted portion of the plate pulls at the other end, i.e., "slab pull." Hence, subduction or plate tectonics, in general, can be considered for the most part a self-accentuating process. Characteristic for convergent plate boundaries are volcanic arcs above the subduction zone and fore-arc deep-sea trenches.

(F) If rates of convergence are greater than those of divergence, the ocean basin begins to close and seafloor spreading eventually ceases.

(G) In the final stage of the Wilson Cycle, a continent–continent collision is the result of the complete subduction of the old ocean basin. During the collision of continents, basement and cover rocks are folded, sheared and thrust over each other, creating a stack of individual units or nappes. The great overburden created by such nappe stacking, in addition to the lateral stress of the converging continents, leads to temperature- and pressure-induced metamorphism (Greek metamorphóo = redesign). Deformational processes such as folding, faulting, and thrusting that occur during orogenesis are largely submarine. The mountain range is uplifted due to isostatic compensation of the greatly thickened crustal stack in the denser asthenosphere. Isostatic compensation is a very slow process; some high mountains like the Alps have not yet reached a state of isostatic equilibrium and are hence still being uplifted. Once erosion overrides uplift, however, the mountain range begins its decline; its peaks are weathered and eroded and the foreland basins are filled with hundreds to thousands meters of sediment. What were once high mountains are eroded to smooth hills over time. Once erosion has erased all topography, the final trace of orogenesis is the so-called geosuture (from the Latin suer = to sew), the scar which marks the former collision zone. The Alpine suture, for example, lies for the most part beneath the Mediterranean Sea and the Ural Mountains mark the collision zone between Europe and Asia.

1.4 Climate and High Mountains

Alongside endogenic processes, climate plays a decisive role in sculpting high mountains, as it does in all other terrestrial landscapes. However, the effect of climatic elements such as insolation, temperature, and precipitation has a much greater impact on the processes of weathering and erosion in mountainous regions than they do on low-lying ones. Extensive talus slopes for example are the result of freeze–thaw weathering, which is exacerbated by the extreme day and night temperature fluctuations above timberline. Torrential rainfall can lead to mud- and landslides that often have catastrophic consequences in the valleys (Fig. 1.13).

The great relief of mountain ranges such as the Himalaya or Karakorum leads to an even greater effect of existing climatic elements on the landscape. Incised valleys of the southern Himalayan slopes extend for great distances with seemingly vertical walls. These impressive gorges are the result of the combination of monsoon rains with the region's great relief energy. The extreme relief causes a high potential energy of water, landslides, or talus blocks and, hence, erosional force greater than that of the Pyrenees or Rockies.

1.4.1 Leeward and Windward Effects

The microclimates created by the high mountains can have an influence of great spatial extent on surrounding landscapes; temperature and precipitation can vary greatly in comparison to regions of a similar latitude and elevation. The wind- and leeside effects for example reach far beyond the foot of the mountain. A remarkable leeside phenomenon is the foehn (adapted in German from the Latin word favonius = warming), a katabatic wind that results from the adiabatic warming of an air mass that has lost its moisture on its way up the windward side. The best-known example is the alpine foehn which usually refers to a foehn wind on the northern slopes of the Alps. Hot moist air masses are pushed up against the Alps from the south. As the air rises, it initially cools adiabatically at a rate of approximately 1 °C/100 m of elevation, until clouds begin to form. As precipitation forms, condensation warmth

Fig. 1.13 Weathering and erosion. Geislerspitzen, South Tyrol, Italy. Extensive talus slopes at the foot of a wall are witness to intense weathering processes above timberline. Published with kind permission of © Dr. Alexander Stahr 2014. All Rights Reserved

Fig. 1.14 Foehn weather. Karwendel Range, Northern Limestone Alps, Germany. The late autumn foehn winds provide pleasant hiking conditions. Published with kind permission of © Dr. Alexander Stahr 2014. All Rights Reserved

reduces cooling to 0.5 °C/100 m. Once all moisture is rained out, the cold, dry air passes the summit and sinks down the leeside — warming adiabatically at a rate of 1 °C/100 m. In this manner air on the leeward side is much warmer and drier than air on the windward side at the same elevation. The foehn can have disastrous effects during the ski season with temperatures rising up as much as 20 °C down the leeward side. Rapid melting often results in avalanches, causing extreme soil erosion especially above the timberline. Extensive and curious erosional forms are witness to such processes. The foehn therefore also has a geomorphologic significance that should not be forgotten. A reverse of winds leads to a foehn wind on the southern slopes of the Alps and the correspondent rainout on the northern slopes — the cold long-lasting rain, prolonged "Schnürlregen" — has ruined countless summer vacations (Fig. 1.14).

The effects of foehn winds are much greater for example in the monsoonal high mountain climate of the Himalaya. The timberline lies at 3,600–3,800 m on the windward slopes of the range. In contrast, the timberline reaches as high as 4,400 m on the leeward slopes, which is the thermal consequence of descending air masses. For the same reason,

Fig. 1.15 Großglockner (3,798 m), Hohe Tauern, Austria. Cool glacier winds make their way from the "eternal ice" down the valleys where their cooling effect shortens vegetation periods. Published with kind permission of © Dr. Alexander Stahr 2014. All Rights Reserved

glaciers are scarce on Himalayan north faces due to higher temperatures and lesser precipitation. The lack of precipitation (and therefore fewer glaciers) reduces erosion rates, which in turn decreases the extent and intensity of geomorphological development. Foehn winds can be found in regions all across the globe; in the Rockies they are called Chinook Winds, Puelche in Chile, or Zonda in Argentina.

1.4.2 Local Winds

Apart from the large scale of leeward and windward wind effects, anabatic and katabatic circulation lead to a number of local effects in high mountain areas. Solar radiation heats air masses during the day, creating warm winds that rise up the slopes, cool, and rainout, i.e., an anabatic wind. At night the cold and heavy (and now dry) air masses flow down the slopes into the valley, i.e., a katabatic wind. In tropical and subtropical mountain ranges local winds greatly influence cloud cover and precipitation distribution in large surrounding valleys. The dry katabatic winds often pose great agricultural problems.

Glacier winds or gravity winds represent a local wind phenomenon. Cold winds are formed at glaciers as a result of the endothermic nature of melting and evaporation of snow and firn. Denser than the surrounding air, the cold glacier winds descend far into the valley. As a result the vegetation period in many valleys is drastically shortened, due to the influx of cold air (Fig. 1.15).

1.4.3 Temperature and Precipitation

In general, elevation gain coincides with temperature decrease and therefore also with a decrease in the duration of alpine vegetation periods. The temperature at high altitude on a mountain, however, will always be higher than an equivalent altitude in the open atmosphere. This is due to the fact that the solar rays reflected by the mountain have a considerable thermal effect on the surrounding atmosphere. Depending on the height and size of a mountain or entire range, warmth can be reflected into great heights. A solitary volcano for example is not capable of reflecting as much heat as an entire range and the extremities of a range will obviously not reflect as much heat as the interior will. This phenomenon is due to the so-called Massenerhebung effect (German for mountain mass elevation). The climatic snow boundary on the northern border of the Alps for example lies at 2,500 m, whereas in the Central Alps it is at 3,000–3,200 m. Annual temperatures at the highest points of the Central Alps are 5 °C higher than the same atmospheric elevation surrounding the Alps.

However, an exception to the general elevation gain is that it coincides with temperature drop rule. Especially in late fall and winter when the sun is lowest, many interior valleys and basins are darkened by the shade of the high peaks almost all day long. The sun cannot warm the air in these lower elevations and the result is a so-called temperature inversion. The valley is covered by a blanket of cold air, while higher up the mountain temperatures rise considerably. Valleys with thick cloud cover and the peaks above jutting into sunny skies are the result of such conditions. South facing slopes can have summerlike weather conditions while the temperature in the interior valleys can be below freezing.

In many high mountain regions that are climatically influenced by eastbound cyclones, precipitation increases all the way to the summit region. This is the case in the Alps, as well as in the Karakorum. In contrast, precipitation in mountain ranges influenced by tropical monsoon systems decreases after reaching a maximum zone. In the southern regions of the Himalaya, summer monsoon rainout leads to 3,000–6,000 mm of annual precipitation from 1,500 to 3,000 m. From 3,000 to 4,000 m less than 1,000 mm is registered per year and above 5,000 m precipitation values drop to 300–500 mm. This is due to the relatively thin (2–3 km) airstream that the southern Asian summer monsoon brings with it. Should the highest annual rates of precipitation occur on the southern slopes of the Himalaya, the interior will be quite dry. Semiarid microclimates are quite often in the interior valleys of the Himalaya.

1.4.4 Exposure

The climate of high mountains is characterized by pronounced differences in exposure, the only exception being arctic ranges in times of midnight sun. Central alpine faces and slopes above the climatic snow boundary that are exposed to the north, for example, often display extensive glaciation — no matter how steep they might be. Flanks with southern exposure on the other hand are often largely free of glaciation, even in the summit region. This phenomenon applies to all ranges of the northern hemisphere, whereas in the southern, the exact opposite applies.

In the winter months, many glaciated central alpine north faces appear almost bare, in contrast to southern flanks buried in snow. The northern flanks of high alpine peaks lack snow in the winter, but are frequently covered by new snow in the summer. The snow cover on the southern flanks vanishes almost completely in the summer months due to increased insolation. In the summer, the northern flanks can still be just cold enough (just below freezing) for wet, sticky snow to accumulate. In the winter on the other hand, cold dry snow that falls on the shady northern flanks does not stick and is whisked away by strong winds. For the most part, it is simply too cold at such heights for the precipitation of cohesive snow. This is especially the case in the high peaks of the Himalaya, where precipitation diminishes rapidly with increasing elevation. Nanga Parbat (8,126 m) and Cho Oyu (8,153 m), however, present an exception; extensive glaciation reaching up to the summit region is the result of a localized climatic phenomenon. Another exception is the small niche glacier that clings to the north face of Mount Everest; dry wind-blow snow in the Norton-Couloir has been compressed into ice over time.

1.4.5 Altitudinal Zonation

The decrease in temperature with elevation (and the reduction of vegetation periods with which it coincides) changes in precipitation altitude or the absolute increase in solar radiation with elevation are some of the many effects of altitudinal zonation. As discussed in Sect. 1.2, a climatically defined zonation or vertical progression of different ecological conditions is the criterion that defines "high mountain" or "alpine." Each zone is subject to different geomorphologic processes and different vegetation conditions. In this right, one speaks of geomorphologic altitudinal zonation and altitudinal vegetation zonation.

The nival or glacial level above the climatic snow boundary is well known from the Alps. Plants are seldom; only specialized species exist in particularly advantageous locations. The next level down, the sub-nival, is characterized by frequent frost. The alpine and subalpine levels follow, which are succeeded by the montane level. At last, the lowland level shows little vegetational difference to the surrounding planes. This altitudinal zonation was derived from the Alps and is often incoherent with mountain ranges outside of central Europe; different climate belts can purvey different vegetation at identical elevations. Furthermore, not all mountain people of the world can necessarily identify with adjectives such as "alpine" or "subalpine." In this right, the mountaineers of South America refer to themselves not as "Alpinists," but of course as "Andinistas."

In South American mountain ranges of the tropics, a different classification of altitudinal zonation is used, which can also be applied to other tropical ranges. The highest level above the climatic snow boundary is called the tierra nevada (=snow land). The following tierra helada (= icey land) is the equivalent of the alpine sub-nival. The next level, the tierra fria (= cold land), is subject to frequent nocturnal frost and is the counterpart of the alpine level. The tierra templada (= temperate land) is found below the absolute frost line and can be compared to the lower subalpine level. The tierra caliente (hot land) constitutes the lowest level with rainforest and savannah (Fig. 1.16).

The globally transferable and therefore less problematic geomorphologic altitudinal zonation can be broken down into four units. The lowermost level extends from the valley floor up to timberline. Geomorphologic evolution in this level is primarily governed by flowing water. The periglacial level follows, characterized by permafrost of considerable penetration depth and frequent surface frosting. Freeze–thaw weathering and solifluction sculpt the periglacial landscape;

Fig. 1.16 Tierra Nevada. Cerro Tapado (5,500 m), Chile. The tierra nevada level is the counterpart to the Alpine nival level. Published with kind permission of © Prof. Dr. Heinz Veit 2014. All Rights Reserved

Fig. 1.17 Permafrost. Furka Pass, Switzerland. Solifluction lobes form above frozen subsurface horizons in the warmer months, as a mass wasting. Published with kind permission of © Dr. Alexander Stahr 2014. All Rights Reserved

solifluction lobes and talus slopes are common (Fig. 1.17). The diversity of geomorphology in the following glacial level is endless. The highest geomorphological level follows the glacial and can only be found in the summit regions of the worlds' highest mountains as well as on the 5,140 m Winson Massif of the Sentinel Range in Antarctica. Weathering in this level occurs almost always well below zero, with daily temperature fluctuations between −10 and −40 °C. Such great temperature fluctuations cause thermal stress weathering, causing a rock's mineral constituents to expand and contract. The rock fragments along grain boundaries are a result of the repeated volumetric strain, leaving only rubble behind.

In the same manner in which the highest geomorphologic level can only be found on earth's highest mountains, other levels can be absent on many high mountains. Dessert mountains like the 3,400 m central Saharan Mount Tibesti or volcanoes like the Hawaiian Islands often lack a glacial level.

1.4.6 Climate Change

The borders of geomorphological as well as vegetational altitudinal vegetation are not always well defined, overlap is common. Borders in the interior of a range are higher than in peripheral areas and this discrepancy increases the higher the mountains are. Differences in exposure can in turn displace boundaries in relation to each other; the altitudinal boundaries by no means conform. The smallest climatic changes can lead to the displacement of zonation boundaries. In the Central Alps, for example, evidence of pedogenesis can be found in connection with solifluction lobes in the periglacial level, which indicates a reprieve in active landscape shaping as the result of a warmer climate. Alpine soil archives reveal evidence for repeated shifts of altitudinal zonation in the order of many hundred meters since the end of the last ice age 10,000 years ago.

The present trend towards a warmer global climate is problematic, especially for the densely populated high mountain regions of the planet. Talus slopes of the periglacial and glacial levels are stabilized by permafrost, the global retreat of glaciers since the mid-nineteenth century has led to extensive subsurface thawing. Catastrophic landslides of unfathomable destructive potential are an unavoidable result of high mountain climate change. Furthermore, glacial shrinkage not only poses great problems for mountain tourism, but also threatens the water supply of millions.

2 Landforms Determined by Tectonics

A. Stahr and E. Langenscheidt, *Landforms of High Mountains*,
DOI 10.1007/978-3-642-53715-8_2, © Springer-Verlag Berlin Heidelberg 2015

Fig. 2.1 Synsedimentary folds. These folds in the Berchtesgaden Alps (Germany) are the result of sliding of the not yet completely consolidated sedimentary deposits on the ocean floor. Published with kind permission of © Dr. Ewald Langenscheidt 2014. All Rights Reserved

Anyone who has ever been in high mountains knows the striking structures in the rocks: folds of varying size and shape. Sometimes whole mountain ranges look like pushed together as a tablecloth. This sample shows very accurately the cause of rock folding. In most cases folding occurs by narrowing the earth's crust. Therefore, wherever tectonic plates are colliding earth's largest, more or less folded ranges of high mountains are arising as mountain chains. These are for example: the Alps, the Himalaya, the Andes, the Rocky Mountains, the Caucasus, or the central mountains of Papua New Guinea.

Besides folds or flexures, we encounter in the mountains other prominent structures associated with the earth's crust by tectonical stress. There are fissures with different characteristics and dimensions to also folded or unfolded thrust sheets or nappes. And one thing is striking in many high mountains: The summits of nearby mountains often have almost the same height. Down in the valley we can see slope flattening strips or podestal levels that mark the rests of old valley grounds. All of these phenomena in the high mountains are the result of tectonic processes over millions of years.

2.1 Folds and Nappes

At the introduction of this chapter, the narrowing of the earth's crust was named for the most common cause of folding in rocks. But there are folding structures which were caused by other processes. Thus, sedimentary layers are pushed up above by more or less vertical penetration of the flowable materials from the depth of the rock. Some know this from salty rocks, which form salt domes or so-called diapirs (Greek diapeíro = penetrated). Besides fractures also folds can form such structures. They can also be formed during the ascent of magma which deformates and folds the surrounding rocks nearby. The thus folded layers are finally existent as dents or fake folds. Also, folding structures can be formed during shifting of sediments as synsedimentary folds (from the Greek syn = together and from Latin sedimentum = sediment). Significant amounts of marine sediments had been shifted to the shelf area of the continents. Beginning to slide, folds may be the only result in the partially solidified deposits. Under water, this process can start if the slopes have at least 3–5° inclination. True folds in the rocks are the result of a lateral narrowing. The previously mounted horizontal sedimentary layers had bent. The extent to which this occurs can be very different. Taking a piece of rock, sometimes tiny folding structures can be recognized. In other cases, folds over several meters or even greater distances occur. Many kilometers long folded structures are also not uncommon. The extent of the layer bending is very variable. It depends on the duration of force and the resistance of different layers of sediments or rocks that oppose rock deformation (Fig. 2.1).

In order to accurately describe folds, the different type of folds and their main structures got a terminology by geologists. A bulge is known as saddle or anticline; a syncline or trough means a concavity. For the sides of a fold the term flank or limb is used. An imaginary surface that divides

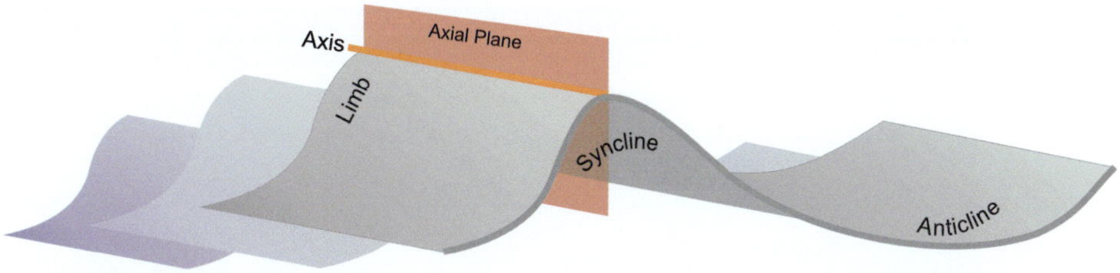

Fig. 2.2 Elements of a fold. Published with kind permission of © Dr. Ewald Langenscheidt 2014. All Rights Reserved

Fig. 2.3 Folds. Eleonorebay, Greenland. Folds in the rocks reflect with few exceptions the narrowing of the earth's crust by lateral compression. Published with kind permission of © Herbert Funk 2014. All Rights Reserved

the two sides of a fold as symmetrically as possible is called crest plane. The line on which the sediment layers turn around is the fold axis. The highest line, where both limbs meet, is called crest. But there are a lot more terms which describe the fold types like asymmetrical, isoclinal, or monoclinal; the folds may look parallel, similar, or like a concertina. Also special terms for the dipping of folds are used (Fig. 2.2).

Not all lines have a vertical axial plane. With increased horizontal force to the folds they can bend more and more. Here the rock strata of the fold limbs are overturned. This finally results in reverse age sequences of the original layers. The primary age sequence of the deposits was thus reversed. The process of folding can go so far that the folded rocks are breaking off and being deported; it comes with faults and thrust faults. The displaced folds or bigger nappes can be transported at high pressure action over considerable distances and may be folded even further. Some speak of thrust sheets or nappes in the Alps or the Himalaya. The motion and thrust of rock sequences and nappes follow not only the lateral thrust, but sometimes according to the slope conditions, it can become independent and then follow the gravity. The reason why nappes exist is the over thrust without significant folding. Examples can be found in the Northern Limestone Alps and the Dolomites.

Under high pressure from all sides, rocks can be deformed plastically, more or less strongly, depending on their facies and mineral composition. Such conditions are particularly given at greater depths below the surface. The condition of a plastic deformation is the relatively high temperature in relation to the surface. The intense folding of the Central Alps and other high mountain areas of the Earth had been formed in several thousand meters of depths as a result of plate collision due to narrowing of earth's sheets. At the same time, the rocks altered under high pressure and temperature; they became metamorphic rocks. Their mineral content adjusted to the new conditions, which is completely different from their original. The process of metamorphism starts at about 200 °C. This corresponds with depths of about 7–10 km in the Earth. In the course of folding caused by directed pressure minerals can newly form parallel textures and thus an entirely different looking rock had been shaped. A sediment such as clay changes to clay or mica schist, limestone becomes marble, and igneous rocks such as basalt or serpentinite will change to prasinite. Climbers know the unpleasant characteristics of metamorphic rocks with a strong schistosity. Handles break easily and safety catches such as hooks or clamping wedges are often difficult to install (Figs. 2.3, 2.4, 2.5, and 2.6).

Fig. 2.4 Wide-area folds: Nappes. Säntis (2,502 m), Switzerland. Folds in the strata from in the Upper Jurassic to Lower Tertiary (Paleogene) of the helvetic zone at Mount Säntis. Published with kind permission of © Dr. Alexander Stahr 2014. All Rights Reserved

Fig. 2.5 Cleveage. Directed pressure may lead to foliation and cleavage of rocks. In contrast to the primary rock stratification as found for example in the Northern Limestone Alps and the Dolomites, tectonic processes and metamorphosis are responsible for the fine and finest platy minerals parallel to the axial planes of the folds. The good cleavage is based on partial recrystallisation of the platy mica minerals, there are often mica flakes in the microscopic field. Mountains which had been built just of slate or mica schists are not prestigious climbing objects. Form and type of cleavage depends on the rock mineral composition, tectonic stress conditions, the temperature and the strength of tectonic deformation. Published with kind permission of © Dr. Alexander Stahr 2014. All Rights Reserved

2.2 Fractures and Related Structures

Fold points, with certain exceptions, are mostly compressed due to lateral forces. In contrast, fractures or faults can arise both through compression and by stretching or shearing of the rock. Like folds fractures also occur in all sizes. Fractures are discontinuities in the rock because of shifts occurred in the past or are still taking place. They are also called dislocations or shifting thrusts. Not very open rock joints, where there has been no major tectonic movement, are called fissures.

Horizontal displacements of rock are known as horizontal or slip faults. They can reach enormous sizes at the edges of tectonic plates. Vertical displacements are named depending on the movement of normal or reverse faults. A stair-like sequence of vertical fractures is called dip-slip faults. It occurs in the landscape morphology in a sequence. The extent of vertical displacement along the fractures is called the height. Deportation fractures or thrusts occur wherever there is a stretching in the earth's crust, nowadays or in the past. The result of relative shifting between two opposing rock fragments is called a flexure (from the Latin flexura = curvature), when there are no significant fractures but only a bending of layers is found (Fig. 2.7).

The movement of the earth's tectonic plates is of course the formation of connected trenches of various dimensions. They vary in size, for example, look at the Upper Rhine Valley, the East African Rift Valley, or the Red Sea (Figs. 2.8 and 2.9).

2.3 Reliefs of Former Times

Looking around from a mountain peak it is striking that many neighboring mountains and massifs have approximately same height. This phenomenon is called the summit plane. The summit plane is a result of landscape development and associated with forming of older reliefs in the Alps. But in many other high mountain areas it is determined and the summit plane corresponds with the highest plane in a succession of flat landscape surfaces in certain height levels. Downhill follow under the recent summit plain younger flat widespread areas. These are characterized in the Northern Limestone Alps and the Dolomites by mountains or massifs with a striking plateau character. For example there are in the Northern Limestone Alps the Steinernes Meer, Hagengebirge, and Tennengebirge. Over millions of years by phases of tectonic uplift and tectonic calm during the whole uplift of the Alps, times of deepening with cutting in rivers and the development of different flat landscape surfaces changed, as it can be seen until nowadays. A storey-like buildup in geomorphological sense was done with the oldest landscape at the top. This shows the sequence

Fig. 2.6 Sliding masses. Untersberg, Berchtesgaden Alps, Germany. By the sliding of large rock masses ("Berchtesgaden nappe") to deeper basin areas folding structures were formed on their forehead. Published with kind permission of © Dr. Ewald Langenscheidt 2014. All Rights Reserved

Fig. 2.7 Alpine fault. The Alpine Fault is a right side oriented transform fault, which runs through the South Island of New Zealand. There the Australian and Pacific plate passes each other. As a result, earthquakes with a magnitude of 7–8 occur relatively frequent and earthquakes with a magnitude greater than 8 about every 100–300 years. To document the horizontal displacement of the two plates is quite simply, a wall was built across the plate boundaries. Published with kind permission of © Dr. Ewald Langenscheidt 2014. All Rights Reserved

of high altitude and steep flat surfaces very clearly to a phased uplift of the mountains.

Left in the valleys, as a result of youngest uplifts and the deepening of glaciers, terraces and slope flattening give the relief in the deeper parts a steplike character. The remnants of the former valley grounds often rise higher platforms above the present valley bottom. Not infrequently they wear settlements, which were protected from the floods of the now deeply incised mountain rivers. On a trip through an Alpine valley, such as the Ötztal in the Central Alps, a number of these old valley bottoms at the valley flank can be observed. Such valley bottoms follow the already existing

Fig. 2.8 Dip-slip faults. The western part of the Hagengebirge, Berchtesgaden Alps, Germany. Fractures are the result of strain, compression or gliding, acting on the rocks of the Earth's crust. The movement of the tectonic plates is because of this force. During the early separation of two plates the plate edges starts to deepen, which is accompanied by fractures in the form of vertical shift faults. The fracture zone here with its stair-like relief was established long before the uplift of the Alps, in connection with the development of an Upper Jurassic basin. Published with kind permission of © Dr. Ewald Langenscheidt 2014. All Rights Reserved

Fig. 2.9 Trench. Mount Watzmann, Berchtesgaden Alps, Germany. A tectonic trench is a sunken piece of crust that is bounded by approximately parallel faults to each side. The scale of trenches can be very different. A huge trench in the Earth's crust can be found for example in the Red Sea or the Upper Rhine Trench as part of a system, which stretches from the Mediterranean near Marseilles, France until Lake Mjösa nearby Oslo, Norway. The picture shows the Eisgraben below the east wall of Mount Watzmann, which is with 1,800 m height the biggest wall of the Eastern Alps. Published with kind permission of © Dr. Ewald Langenscheidt 2014. All Rights Reserved

valley, while the wide ranging elevated plateaus of former times above cross over today's watersheds and faults and form other structures.

Up to the recent valleys and valley grounds all relief generations existed even before the Pleistocene glacial periods, which began about two million years ago. High situated hang up bars or flat surfaces formed as remnants of the former land surface during the ice ages and were starting points of the glaciation. The ice streams dissected individual mountains from each other and also striking

Fig. 2.10 Summit plain. It is often observed that the peaks of high mountains, such as the Himalayas or the Alps, are over long stretches in the same level, whatever the mountain structure and the different rocks may be that build up the mountains. That is the so-called summit plain which is the result of tectonic uplift of a formerly shallow, flatly relief. In the Alps several times phases of fast uplifts changed with longer phases of landscape forming. The result is now a storied building of the alpine landscape. The summit plain is the highest remainder there. View from the Chiemgau Alps to the southwards Central Alps, in the middle of the photo the Großvenediger, 3,662 m, in the Hohe Tauern. Published with kind permission of © Dr. Alexander Stahr 2014. All Rights Reserved

Fig. 2.11 Old valley grounds. Dolomites, Italy. The phase of uplift of young mountains led to the valley bottoms were cut and their sediments moved away. The remains of the old valley grounds (*middle* and *left*) can be found today in the form of strips or terraces high above the valley ground from nowadays. Published with kind permission of © Dr. Alexander Stahr 2014. All Rights Reserved

cirques (conical wall niches) were formed. Hang strips or old valley floors at the upper boundary of the ice age glaciation became altered.

From the central mountain range to the outer zones of the Alps, in the Himalaya and other mountains can also be recognized a stair-like sequence of levels of peaks of equal height, due to an earlier uplift and higher elevation or to dome-shaped bulge of the inner mountain ranges. In the best-studied high mountains of the Earth, the Alps, the legendary geomorphological sense (from the Greek gé = ground, form and lógos morphé = science, teaching) was already recognized at the turn from nineteenth to twentieth century and has been repeatedly the subject of research since then.

Particular attention fell on the “Augensteine”; the word means “Pebbles looking like eyes,” which are smooth polished river pebbles from weathering-resistant rocks.

Fig. 2.12 Peak levels. Steinernes Meer, Salzburg Limestone Alps, Austria. A descending height from peak levels from the central Alpine area to its edges become clearly as a result of their greater uplift during mountain building. Published with kind permission of

Fig. 2.13 Former landscapes. The Steinerne Meer and the Hagengebirge (*photo at the center*) are a prime example of a former landscape, which was active in the course of the development of the Alps during Paleogen and Neogen. In that time this plateau of the mountain range was part of a shallow, undulating relief. From the territory of the already more uplifted Central Alps river systems transported their gravels to the north and deposited it. During the subsequent uplifts, the relief remained largely preserved. First protected by the crystalline gravel, and then, after their erosion, the karst plateau of the mountain preserved the former landscape. While during the phases of uplift the surrounding river systems cut deeper and deeper, the drainage of the plateau as of now went underground. The gravels which exist only there in isolated and several times rearranged remnants tell us about the tectonical and geomorphological history of the mountains here. Published with kind permission of

These are common in the Northern Limestone Alps on the flat high relief landscape surfaces as residual sediments. This allochthonous Augensteine comes from the Central Alps and shows south–north beds of river gravel, due to the greater uplift of the Central Alps at a time in which the Northern Limestone Alps were a low undulating plateau landscape with a northern directed slope erosion surface.

The gravels of the Augenstein landscape — this term is often synonymous with the term Raxlandschaft — were cleared up to today and have already relocated due to erosion and the occurrence of pleistocene glaciation. The fact that the plateaus of the Steinerne Meer, the Hagengebirge, and other regions of the Northern Limestone Alps were not strongly eroded after the erosion of the outer layers is mainly due to the karstification of these areas, a difference to the flat surfaces of the Central Alps which had been eroded stronger. While outside the plateaus flows worked at the depth of erosion, the drainage in the karst areas was largely underground. With the lowering of the surrounding valley areas through increased erosion due to tectonic uplift, a lowering of the karst water table was connected. It could develop a number of cave systems within different levels of height also, each corresponding to a landscape surface. So the cave systems are in strong relationship to the former landscapes.

The development of many former high mountain reliefs is also evident from the fluvial sediments transported into the foreland. In periods of tectonic calm, which led to the development of flat landscape surfaces, mostly fine sediments such as clays and sands were transported by rivers to the basin in front of the Alps. On the other hand the coarse of bed gravels often represents phases of tectonic uplift and increased erosion (Figs. 2.10, 2.11, 2.12 and 2.13).

3 Volcanoes

A. Stahr and E. Langenscheidt, *Landforms of High Mountains*,
DOI 10.1007/978-3-642-53715-8_3, © Springer-Verlag Berlin Heidelberg 2015

Fig. 3.1 Lava. Mauna Loa (4,170 m), Hawaii, USA. Molten rock is erupted onto the surface as lava from the volcano. Published with kind permission of © U.S. Geological Survey (USGS, public domain) 2014. All Rights Reserved

Within high mountains such as the North American Cascades or the Andes are numerous peaks of volcanic origin. Molten rock rises from the interior of the earth upward, as it is less dense than the surrounding rock. It forms within the lithosphere magma chambers and from there it moves to surface. This process is called volcanism and includes also the expulsion of gaseous substances at the surface. A volcano is generally the place where these substances reach the surface. This may be at a mid-ocean ridge or in a big ditch like the East-African Rift Valley. The best-known manifestations of volcanism are the cone or shield volcanic mountains. They can reach heights of several thousand meters and often accommodate a pronounced glaciation. The word "volcano" derived from the island "Vulcano" north of Sicily. Because of frequent eruptions on this island, the Romans attributed their existence to the fire god "Vulcanus" (Fig. 3.1).

3.1 The Formation of Volcanic Mountains

Volcanic mountains are caused by the conveying of magma. Magma is molten rock within the earth. If it is transported by a volcano onto the surface, it is called lava. Fresh lavas have melting temperatures of about 1,000 °C. Depending on the chemical composition of lava the temperature varies between 800 °C and 1,200 °C. Such high temperatures which lead to rock liquefaction occasionally occur at depths of 30–40 km, i.e., in the lithosphere. Regularly they are reached at 75–250 km depth in the asthenosphere.

Due to the enormous temperature and pressure exerted on the asthenosphere, it is not liquid, but plastic. It can therefore be regarded as relatively fixed because pressure acts against liquefaction. The melting and formation of magma are usually only a result of anomalies in the prevailing pressure and temperature conditions, i.e., where temperature is very high relative to the pressure or where the compression imposed on the hot rock masses is reduced. We find these anomalies where tectonic plates diverge and where one plate plunges under another and in the vicinity of hot spots. By convection plastic material rises up along the line of separation between two diverging plates. The lowered pressure at this area leads to melting of the plastic material.

Where one tectonic plate plunges under another, the pressure–temperature conditions are disturbed in the asthenosphere. In addition, water-rich sediments from the ocean floor are drawn into the depths during the collision of an oceanic plate with a continental plate, so that the melting temperatures of rocks within the continental plate are reduced. For these reasons, volcanoes occur in certain zones and belts of the earth, usually at the boundaries between tectonic plates. The exceptions are the hot spots which can occur within tectonic plates.

Some minerals in the earth's mantle melt at lower temperatures than others. The magma resulting from mantle melting is therefore a viscous fluid mixture of both liquid and hot but solid crystals. The geologist describes this physical state as partially molten. The physical properties of magma are thus connected to mineral separation chemistry, melting point, and density — each temperature range representing a particular partial melt.

Geothermal Energy

Our correct understanding is that geothermal energy has three causes: some of the geothermal energy is a remnant from the early Earth. Secondly, during the repeated impact of cosmic material, much of the impactor's kinetic energy was converted into heat. With the accumulation of new material, the mass of the earth increased, raising the gravitational pressure in its interior, causing a rise in temperature. The third and probably the most important process that provides permanent geothermal energy is the decay of radioactive elements such as uranium or thorium. These atoms spontaneously decay by emitting an electron or alpha particle. The released particles are absorbed by the surrounding material and their kinetic energy is converted into heat in sufficient quantity (over billions of years) to partially melt the rocks.

The increase in temperature with depth inside the earth is called the geothermal gradient. On average, the temperature increases by about 3 °C per 100 m in depth. However, deviations occur from this average. They depend on the local geological structure of the lithosphere, the thermal conductivity of rocks, the rocks' bearing, the morphology of the earth's surface, and the occurrence of magma-derived heat sources. In some places within the earth's interior, the temperature reaches 1,000 °C at 30–40 km depth, whereas at others such depths temperatures of only around 500 °C exist. From the upper boundary of the mantle to its lower boundary at 2,900 km depth, temperatures range from about 1,000 to 2,500 °C.

Due to the very complex process of magmatic differentiation various different magmas can be formed at different temperatures. Most of the magma formed in this way is of basaltic composition. Molten rock has a lower density than solid rock and therefore ascends (as oil does in water) towards the surface. It rises through pores and fractures in the overlying bedrock and can form magma nests. Finally, the surrounding rock is melted or pushed aside, generating large magma chambers which can occur at different depths within the lithosphere. Their dimensions can grow to several kilometers. From there, magma rises as a result of accumulating pressure through columns or volcanic vents up to the surface and becomes lava following a volcanic eruption. In very little time volcanic mountains may be constructed, which can grow to enormous heights. One of them rises 4,202 m above sea level — Mauna Kea of Hawaii. With an overall height (from the bottom of the ocean to its peak) of 10,203 m, it is undoubtedly the highest mountain in the world. As to its origin, Mauna Kea is atypical (like the other Hawaiian volcanoes) since it owes its existence to a hot spot and is surrounded by a single lithospheric plate (Figs. 3.2, 3.3, 3.4, and 3.5).

Hot Spots

Hot spots are volcanic manifestations constructed by a so-called mantle plume, consisting of a narrow jet of hot material rising in a columnar-like fashion (due to its lower density) from the interior of the earth's mantle. The magma melts the oceanic plate at a point, forming a magma chamber which penetrates into the lithosphere along vents, ultimately flowing out onto the seafloor. The hot spot jet does not change its position with respect to the earth's mantle. If a plate above a hot spot is tectonically moved laterally, it leaves behind a series of extinct volcanoes of increasing age. The island chain of the Hawaiian archipelago is such an age series. Hawaii (the youngest island) began to form on the ocean floor about two million years ago. Midway to the northwest is about 30 million years old. The volcanic islands of the Hawaiian archipelago, with peaks more than 4,000 m high, are situated in the middle of the Pacific plate, a fact which was difficult to explain using the theory of plate tectonics. A stationary hot spot did not seem a real weld in the lithosphere moving over it, because the escaping lava does not come up as a continuous stream from depths, but at certain intervals as a mantle plume. Even the volcanism of the Cape Verde Islands, the Galapagos Islands, or the Azores is attributed to hot spots.

3.2 The Distribution of Volcanic Mountains

Most of the volcanic mountains occur in conspicuous belts, chains, or narrow lines. These mark the boundaries between tectonic plates and are also regions of increased seismic activity. At locations on earth where tectonic plates move away from each other, volcanoes fill the resulting gaps between drifting plates with basaltic lava flows. Thus, mid-ocean ridges occur and give rise to new seafloor. This form of volcanism is known as rift volcanism and it continuously creates new seafloor material, while older seafloor is swallowed by subduction zones. Although there have been oceans for billions of years, all of the seafloor on Earth owes its existence to this rifting process and is never older than about 200 million years.

The most dangerous volcanoes are those built upon regions where one lithospheric plate plunges under another and descends into the asthenosphere. They are located a few tens of kilometers from the plate boundary on the overlying plate and are formed by partial melting of rocks in the upper part of the subducting plate. Their danger lies in their

Fig. 3.2 Scoria throw-off (Spattering). PuuŌō, Kīlauea, Big Island, Hawaii. If magma with a high gas content rises through the vent of a volcanic mountain, it foams as a result of the pressure release and breaks into vesicular fragments, which are expelled from the volcanic vent. The coarse fragments are referred as scoria. The term "scoria," as well as the term "volcanic ash" is somewhat confusing, as the products of a volcano do not constitute of combustion residues. If a volcano blows out only volcanic slag, this is called a scoria throw-off. During the flight, the material cools down and forms so-called pyroclastics (from the Greek pyr = fire and klásis = to break). Additionally there are also finer particles such as dust and ashes. Even rock debris, which has been excavated from the walls of the vent during an eruption, may be distributed among the pyroclastics. If ejected lava fragments are fused together upon landing, they are called "weld slag." Published with kind permission of © U.S. Geological Survey (USGS, public domain) 2014. All Rights Reserved

Fig. 3.3 Pahoehoe lava. La Palma, Spain. If a basaltic lava flow cools down slowly, a skin is formed on the surface. Under this thin skin the lava flow continues to flow and moves itself ropelike. The Polynesian name pahoehoe means knitted or twisted like a rope. This type of lava is typical for the Hawaiian volcanoes. Due to the low viscosity of the lava it can be transported over long distances and formed enormous volcanic mountains with only gentle slopes. A well known example of such a shield volcano is the Mauna Loa on Big Island, as the locals call Hawaii. Published with kind permission of © Dr. Ewald Langenscheidt 2014. All Rights Reserved

Fig. 3.4 Aa-lava. La Palma, Spain. Aa-lava looks like plowed field soil. This lava has largely lost its gaseous constituents, making it tougher. During cooling down it forms a thick crust that shatters with further movement into rough, sharp-edged boulders and blocks. The polynesian name is very revealing. Aa is the word that you would pronounce if you traverse this lava barefoot. Published with kind permission of

Fig. 3.5 Volcanic ash. Ash arises when an explosive eruption takes place and is also called tephra (Greek tephra = ash). The term is technically incorrect, since ash is not a product of combustion. Volcanic ash is a major threat to aviation, as it can reach up to several thousand meters in height. It can lead, among other things, to airplane engine failures. Published with kind permission of

explosive nature. During the descent of an oceanic plate, water-rich sediments from the seafloor are pulled down in the subduction zone. The water evaporates and advances overheated explosive activity of volcanoes. Moreover the melting of the subducted rocks is assisted by the water. They can melt earlier (at lower temperature) and consequently they rise due to their lower density than the cold rocks above and form volcanic series parallel to the plate

Fig. 3.6 Mount St. Helens. In 1980 Mount St. Helens erupted, located in the North American Cascade Range. The eruption was so violent that it destroyed the northern flank of the mountain and 57 people died. Published with kind permission of © U.S. Geological Survey (USGS, public domain) 2014. All Rights Reserved

margin. A major contrast with the volcanism of mid-ocean ridges is the occurrence of various types of volcanic rocks. Within the subducting plate are basaltic crust and ocean floor sediments, which are partially molten to form andesitic magma and because of its high viscosity it causes explosive eruptions (Fig. 3.6).

Where an oceanic plate plunges under another oceanic plate, volcanic island arcs are formed as in the case of the Philippines. If an oceanic plate moves under a continental plate, volcanic mountain chains arise such as the Andes. When a continent cracks, continental rift vulcanism occurs. A characteristic feature here is the emergence of a rift trench, which is characterized by basaltic volcanism as exemplified by the East African Rift Valley. The rift development represents the early stage of the formation of a new ocean.

Fig. 3.7 Shield volcano. Mauna Loa, Hawaii, USA. Shield volcanoes result from the superimposition of numerous basaltic lava flows. These are very thin and they spread over large areas. Accordingly, the flanks of shield volcanoes are almost flat with inclinations typically under 10°. Mauna Loa (illustrated here) is 4,170 m high and has a base diameter of about 200 km. After the 4,202 m high Mauna Kea, it represents the second highest point of the Hawaiian archipelago. Nowhere on Earth are larger volcanoes than the volcanic mountains on Hawaii. Because the lava at Hawaii usually emerges in low-viscosity lava flows, the Hawaiian volcanoes are considered as meek — in contrast to the often explosive Andean volcanoes. Published with kind permission of © National Aeronautics and Space Administration (NASA, public domain) 2014. All Rights Reserved

3.3 Types of Volcanoes

Geoscientists, specifically the volcanologists, have long tried to classify the different volcanoes and their eruption mechanisms. Volcanic eruptions are affected by many factors, and furthermore, the shape of a volcanic mountain may be the result of various eruptions. Therefore, a classification of volcanoes is always a little arbitrary. Volcanologists classify volcanoes according to the nature of their eruptions and their characteristic properties and by the morphology.

Geomorphologically looking at high mountains, there are essentially two different types of volcanoes: strato-volcanoes and shield volcanoes. Due to their relatively "good-natured"

Fig. 3.8 Strato-volcanoes. Volcanoes which comprise both volcanic tephra (ash, slag) and lava are called strato-volcanoes (from the Latin stratum = blanket, cushion). Other designations for this volcano type are *layer* or *mixed* volcano. The picture shows the 2,518 m high Mount Egmont or Taranaki on the west coast of New Zealand's North Island. It is with an age of 10,000 years the youngest member of the volcanic chain in New Zealand. This typical strato-volcano with its distinctive *conical shape* is often referred to as "New Zealand's Mount Fuji" by many Japanese tourists. Published with kind permission of © Dr. Ewald Langenscheidt 2014. All Rights Reserved

effusive eruptions (from the Latin effundere = to pour out) of low-viscosity basaltic lavas, the volcanologist speaks of *Hawaiian* eruptions, which construct gently sloping dome-shaped mountains, the so-called shield volcanoes. A classic example of a shield volcano is the 4,170 m high Mauna Loa in Hawaii, one of the most powerful and most active volcanoes in the world (Fig. 3.7). The term "shield volcano" was given to this gently sloping volcanic mountain by its resemblance to the "hump shields" of Roman legionnaires.

Alternating phases of effusive and explosive eruptions give rise to strato-volcanoes (from the Latin stratum = blanket, upholstery), such as Mount Egmont or Taranaki in New Zealand (Fig. 3.8). Other popular strato-volcanoes include the 6,310 m high Chimborazo in Ecuador, Mount St. Helens (which following its eruption in 1980 had a height of 2,549 m), or the 3,718 m high Pico del Teide on Tenerife. Strato-volcanoes are the most common form of large volcanic mountains.

4 Weathering and Erosion Form High Mountains

A. Stahr and E. Langenscheidt, *Landforms of High Mountains*,
DOI 10.1007/978-3-642-53715-8_4, © Springer-Verlag Berlin Heidelberg 2015

Weathering and erosion are the essential processes which have formed today's high mountain areas and still continue to change them. High mountains arise through the collision of tectonic plates with the exception of those formed by volcanic activity and they begin to collapse even as they continue to grow. Many mountain hikers and alpinists are surely very familiar with weathering and erosion. Think of rockfall caused by animals, unfreezing processes, a rope team, or the actions of a touring company. And what climber has not sworn in anticipation of the torment of traversing a nearly endless rock fan up to the beginning of the climbing route grounded in solid rock? Occasionally one is surprised by large boulders at the wayside which look as if they were split with a wedge. However, the splitting is nothing more than the result of the difference between the day and night time temperatures which cause fracturing within the rocks. Reports and TV pictures frequently reach us concerning destructive mudflows, rock debris flows, and landslides in the high mountains around the world. To all of these phenomena of mountain regions one thing is common: They are the result of uplift, weathering, and erosion processes.

4.1 Weathering

The term "weathering" summarizes all physical, chemical, and biological processes which lead to the loosening and destruction of solid rocks. By weathering processing principally the rock is changed into three forms of material:

A: Solid weathering products taking the form of mechanically formed rock fragments of differing grain size including lumps of rock, rubble, or sand without any chemical change.

B: Dissolved substances by chemical or biological weathering which are then transported away by water as ions or complex compounds.

C: New chemical complexes like clay minerals and oxides which determine for example the color and chemical quality of soils.

Primarily the processes of weathering are controlled climatically by temperature and precipitation. Mechanical biogenic factors like the root pressure of plants or chemical biogenic processes by secretions of organisms are less important. The most important weathering agent is water. However the chemical composition of the rocks, their mineral structure, their permeability to water, their stratification, beds and fractures, or cleavage plains also play an essential role. Ultimately the chemical and structural properties of a rock determine whether it exhibits low or high resistance against weathering processes.

4.1.1 Physical Weathering

Ordinary physical or mechanical weathering is predominant in the high mountains. It breaks the in situ rocks down to boulders, debris, sand, or dust pieces without effecting chemical changes.

Grain Sizes

Solid rock is fragmented by mechanical weathering into different particle sizes which are assigned to different grain size groups depending on their equivalent diameter. The silt group is defined by diameters between 0.002 and 0.063 mm. Sand contains the grain size range 0.063–2.0 mm. One describes grain sizes between 2.0 and 63 mm (depending on whether it is angular or rounded material) as breeze or gravel. Even larger classes are named cobbles and boulders. All these groups can be differentiated still further, for example, into very fine, medium, coarse, and very coarse. In cooperation with chemical weathering, mechanical weathering may also produce even smaller particles with sizes under 0.002 mm. These are described as clay. A mixture of clay, silt, and sand is called loam which can appear as sandy clayey loam, silty loam, or weakly sandy loam depending on the quota of the individual grain groups. The weathered material broken down mechanically becomes an unconsolidated sediment after deposition and forms surface layers on the in situ rock. It is the subsoil substratum of soil formation. The grain size composition of loose material or the soil developed on this is also described as soil type.

Physical weathering attacks the rock at zones of natural weakness which can be fissures, joints, and/or cracks. A fissure is defined as a fine or incompletely opened cleft. In the case of cracks, the opposed rocks have already moved apart. Fissures and clefts arise through compression, stretching, or shearing of the rock due to tectonic movement. Igneous rocks show very regular crack systems caused by cooling of the molten rock. The crack formation of basalts is typical for example. Cracks arise also when slowly cooling plutonic rocks like granite decrease in volume. However, cracks can also arise if the surface layers of mountains are eroded and load removal starts. These types of cracks are called load removal cracks (Figs. 4.1 and 4.2).

4.1.1.1 Insolation Weathering

Another form of physical weathering is temperature or insolation weathering. It is a result of the fact that all materials

Fig. 4.1 Jointing of rocks. Cassianer Dolomit, Sextener Dolomites, Italy. The weathering starts in the form of frost blowing-up attacks at fissures and joints. One describes the spatial arrangement of joints as a joint net, with joints running approximately parallel in the same direction as the joint set. The photograph shows a distinctive jointing in the dolomite, which can be explained by tectonic stress due to mountain formation. Published with kind permission of © Dr. Alexander Stahr 2014. All Rights Reserved

change their volume with temperature. The daily change in insolation and nightly cooling of the ambient air causes expansion and contraction of the rock, which leads to disruption of its mineral structure. This process is exaggerated by the different amount of thermal expansion exhibited by the rock's constituent minerals. In extreme cases larger rocks can actually "burst." Insolation weathering particularly takes place in desert mountains which experience extreme temperature fluctuations.

4.1.1.2 Frost Splitting

As a rule, frost splitting prevails in most high mountain areas. If water freezes to ice and has penetrated into joints and clefts, it increases its volume by 9 %. As a result it develops an enormous expansive force. It obtains its greatest effect at −22 °C with a resulting force of 2,100 kg/cm^2. The change through freezing and unfreezing gradually loosens the rock structure. Joints and cracks expand little by little. Finally the rock disintegrates into debris or frost rubble with all sorts of grain sizes. The more water-filled pores, joints, and cracks exist, the more intensive this form of physical weathering is in appearance (Fig. 4.3).

4.1.1.3 Salt Weathering

In arid mountains dissolved substances are taken up through capillary action into the finest joints by the sporadic occurrence of water. Salts are deposited in these capillary joints if the water evaporates. The crystallization pressure of the salts remaining behind in the fissures destroys the rock mechanically. Tensions of several 100 kg/cm^2 can form which loosen the rock through the change in hydration and dehydration. The rock weathers by blasting up single center punches, fine rock debris, or whole rock peels (Fig. 4.4).

4.1.1.4 Physical Biological Weathering

The growth pressure of plant roots which squeeze into rock clefts leads to the loosening and blasting of rock fragments (or bigger rock pieces) and is termed physical biological weathering. The loosening of rocks by the work of digging animals also belongs to this category of weathering.

The smaller the parts into which the rock is broken through physical weathering, the more easily they can be removed. The size of the weathered material therefore is closely related to the erosion processes (which may be gravitation or a solvent such as water). Both processes, weathering and erosion, are therefore closely connected and support each other (Fig. 4.5).

4.1.2 Chemical Weathering

Unlike physical weathering it comes to putting into action between rock and water by chemical weathering. By this way of weathering the rock is attacked by water alone or by substances dissolved in the water. For example, the simple solution of halite in water is called a product of solution weathering.

4.1.2.1 Carbonic Acid Weathering

Carbonic acid weathering is an important weathering process in many high mountains formed from limestone rocks. Carbon dioxide from the air dissolves into rainwater to form a weak acid called carbonic acid:

$$H_2O + CO_2 \rightleftharpoons H_2CO_3 \rightleftharpoons H^+ + HCO_3^- \quad \text{(formation of the acid)}$$

Fig. 4.2 Shrinking clefts. Ventilation arises also in volcanic rock. These basalt columns are typical cooling structures. Published with kind permission of © Dr. Alexander Stahr 2014. All Rights Reserved

Fig. 4.3 Frost splitting. Dolomites, Italy. Because of the freezing and thawing of water in rock, frost rubble is formed in different sizes. The roughness of the components depends on the rock type. Finally more or less big rubble tips were formed at the foot of the walls. Published with kind permission of © Dr. Alexander Stahr 2014. All Rights Reserved

While pure water is only able to dissolve insignificant amounts of lime (or more specifically its primary constituent mineral calcite ($CaCO_3$)), the solution of the rock is accelerated when carried out in the presence of CO_2:

$$CaCO_3 + H^+ + HCO_3^- \rightarrow;\leftarrow Ca^{2+} + 2HCO_3^- \quad \text{(lime solution)}$$

Water containing CO_2 etches lime rocks and weathers them. This process is more intensive for purer lime rocks. A high partial pressure of CO_2 in soils also has an effect on lime rock, for which numerous plants and microorganisms are responsible. Manifold land forms, such as clints or lapiés (see below) which can occur even in the steepest slopes, karst tables, or karst caves, are the result of limestone solution.

All landscapes which show patterns of denudation affecting their limestone and dolomitic rocks are said to exhibit karst scenery. Corrosion (Latin corrodere: to eat away) refers to chemical attack on carbonates and is the process

Fig. 4.4 Salt weathering. Red Sea Hills, Egypt. Within desert environments the salt weathering is effective in the mountains. Published with kind permission of © Dr. Ewald Langenscheidt 2014. All Rights Reserved

Fig. 4.5 Growth pressure of plant roots. Roots of plants exert an enormous pressure on rocks and loosen them. Published with kind permission of © Dr. Ewald Langenscheidt 2014. All Rights Reserved

responsible for karst scenery. The word karst was derived from the >1,000 m high Slovenian–Croatian limestone mountains of the same name, which extend between Ljubljana, Trieste, and Rijeka and further to the eastern side of the Adriatic Sea.

The calcium magnesium carbonate compound dolomite $CaMg(CO_3)_2$, known for its abundance within the Dolomite mountains where it builds up giant peaks, is not so easily soluble but just like lime as a bicarbonate. Carbon dioxide is soluble even in cold water, so carbonic acid weathering can take place even in arctic regions.

The process of lime solution can be reversed. Dissolved lime is precipitated from water by releasing CO_2 to the surrounding air. Among other things sinter formations (like dripstones) result which are known as stalactites and stalagmites, building fascinating structures inside karst caves (Fig. 4.6, 4.7, 4.8, 4.9, 4.10, 4.11, and 4.12).

Fig. 4.6 Lapiés (Karren). Berchtesgaden Alps, Germany. Rainwater represents a weak acid due to its carbon dioxide content. The formation of lapiés is therefore dependent on the effect of slightly acidic rain water runoff, affecting a more or less strongly inclined limestone surface. Plump forms usually arise under soil cover and sharp-edged directly at the naked rock surface. In the extreme case so-called roof ridge lapiés are developed on older larger round ones, sometimes closely aligned. So the markedness of the lapiés can point to a former vegetative soil cover. The limestone solution often follows the jointing of the rock, the result being lapiés which are aligned crossways to the dip direction from time to time. Snow thawing slowly causes a collection of small hollow forms on the lime, called niche lapiés. A widespread area with all these surface forms is called a limestone pavement. Published with kind permission of © Dr. Alexander Stahr 2014. All Rights Reserved

Fig. 4.7 Doline (Sink). Reiteralm, Berchtesgaden Alps. The conical depressions arise either from limestone solution or from the collapse of the rock into caverns beneath. Forms of dissolving represent a special type of dolines in salty rocks. Published with kind permission of © Dr. Ewald Langenscheidt 2014. All Rights Reserved

Fig. 4.8 Uvala. Funtensee, Berchtesgaden Alps, Germany. Most uvalas should have arisen from the merging of several dolines. Published with kind permission of © Dr. Ewald Langenscheidt 2014.

4.1.2.2 Oxidation Weathering

Iron-bearing minerals are very common and form as a rule dark colored rocks. Oxygen dissolved in water will attack the iron in these minerals and oxidize it. Divalent iron becomes trivalent iron by losing an electron and as a result the rock develops a brownish, reddish, or yellowish color. In short, it "rusts" and the original mineral structure is destroyed by the oxidation weathering. We know this rusting of iron well, as we are all familiar with iron objects which oxidize through prolonged contact with the damp atmosphere. Cars provide a good example (Fig. 4.13).

Fig. 4.9 Polje. Poljes (Serbo-croat. = field) are the largest closed hollows in karst scenery covering tens of square kilometers. They originate mostly by chemical and mechanical expansion of tectonic weakness zones or from the lateral merging of smaller solution-derived forms like dolines or uvalas. Poljes which have developed without some structural influence are called solution or collapse moulding poljes. Poljes always have an underground drainage. They can be permanently dry or show a periodical or permanent water covering. If their surface is covered by fine-grained sediments, the poljes are used for agriculture (see name!). The photograph shows a large polje situated 1,100 m above sea level in the Lefká-Ori on Crete, the Omalós plateau. Published with kind permission of © Dr. Alexander Stahr 2014.

4.1.2.3 Silicate Weathering

During the process of hydrolysis or silicate weathering (Greek. hýdor = water and lýsis = solution) silicate minerals (Latin. silex = pebble) are attacked by water. To the group of silicate minerals which make up about 60 % of all minerals of the earth's crust belong among others feldspars and mica as components of granites besides quartz. Hydrolysis causes exchange of ions at mineral surfaces and therefore strong material changes up to complete disintegration of single minerals and with that the disintegration of their host rocks. This process is favored by gone ahead hydratation weathering which is based on the accretion of water molecules to mineral intersections and leads to the breaking up of the crystal lattice.

Fig. 4.10 Karst pavement. Hagengebirge, Germany/Austria. The karst pavement of the Hagengebirge is over 50 km^2 wide. Published with kind permission of © Dr. Ewald Langenscheidt 2014. All Rights Reserved

Newly formed minerals such as clay minerals are end products of silicate weathering. They play an important role as an exchanger of ions in soil and therefore for the supply of nutrients to plants. A very well-known clay mineral arising from silicate weathering of the feldspar orthoclase is kaolinite (Mount Kaolin in southwestern China), which forms a white to cream-colored clay and is an important raw material for the porcelain industry.

4.1.2.4 Chemical Biological Weathering

Rocks can also be decomposed by chemical biological processes. Higher plants but also algae, lichens, and mosses, which are in direct contact with the rock, release H^+ ions, which attack the minerals within the rocks and destroy them gradually (Fig. 4.14).

4.2 Erosion

The erosion of high mountains can be carried out in very different ways. The mountains are always a place of predominant erosion and the greater the relief energy, the more intensive the erosion will be in all its forms. It is easily understandable that fall masses, in the form of rockfall, avalanches, or mountain falls, are more effective along higher and steeper slopes. Erosion processes therefore take place in extremely high mountains, like the Himalaya or the Karakorum, to a greater extent and with a considerably more widespread effect than on Svalbard or in the Scandinavian high mountains. At the more than 1,000 m high walls of the Himalaya, gigantic ice avalanches can spread over enormous distances and stress the rocks particularly intensively by virtue of their high kinetic energy. Real wall ravines are

Fig. 4.11 Karst table. The karst tables are a striking feature in karst scenery. They belong to the small forms of the karst pavement. The underlay of the "tabletop" is temporarily protected from stronger corrosion similar as in the case of glacier tables and earth pyramids. The photo shows two typical karst tables in the Steinerne Meer, Salzburg, Austria. The extent of solution of limestone since the last ice age can be determined by the height of the karst tables. A lime erosion rate of 1 cm per 1,000 years has been calculated for the area of the Steinernes Meer. Published with kind permission of © Dr. Ewald Langenscheidt 2014. All Rights Reserved

the result of this kind of erosion. We find comparable wall heights in the Alps only in the Monte Rosa massif, whose east wall is the highest wall of the Alps with 2,400 m of height.

The climatic conditions of a high mountainous region are also decisive for erosion. Together with the high relief energy at the Himalaya southern range, the monsoon precipitations for example cause an intensive, almost ravine like carved valley formation.

Or we all know the signs which warn us of rockfall on mountain roads. In the media it is frequently reported that disastrous mudflows and landslides follow intense rains and that the increasing soil erosion by man is similarly problematic. Many high mountainous regions have to thank erosion by glaciers for their visual appearance today. So it is obvious that erosion can be caused by the effect of gravitation alone or by a medium like water, wind, snow, or ice. All processes which lead to the erosion of rocks are described as mass shifts. Gravitative caused shifts of rock are described as pure mass movements or mass wasting. All kinds of fall, processes of slides, and motions of flow are included. Mass shifts which are facilitated by another medium, whether a glacier or running water, are called bulk transport and the erosion by the transport medium is called "Massenschurf," a German word for this phenomenon.

At the more two-dimensional erosion by mass movements like falls, landslides, or motions of flow as well as by the erosion of glaciers, snow, and avalanches, one also speaks of denudation (Latin denudare = to lay bare). On the other hand stand the processes taking place more linearly, under which the processes caused by runoff waters take in the high mountains an outstanding position. In the following the mass movements and the results of the "massenschurfs" by water erosion, man, and grazing animals will be treated nearly. The "massenschurf" by snow, avalanches, and glaciers is treated in separate chapters as well.

4.2.1 Falls

The fall of single large rock fragments from steep faces is called rockfall. Rockfall usually starts with the morning warming on the day when cleft ice unfreezes between the loosened rocks — ice which had held these blocks together previously. Therefore rockfall is very common out of the dampest wall parts. Firm rains can assist rockfall because water reduces the friction between adjacent rock fragments and therefore causes their fall. The removed debris falls down to the foot of the rock face and forms a rubble tip over the course of time. It consists of all sorts of grain sizes, from coarse blocks and stones up to sand and finest rock particles. For any particular fall the coarser components generally travel furthest downwards since they have a larger mass and consequently a greater kinetic energy. The finer components on the other hand accumulate in the upper area of the rubble tip (Figs. 4.15, 4.16, 4.17, and 4.18).

If the internal cohesion of larger rocks is reduced, rockfall avalanches can be caused. This is in principle nothing but an enormous stone fall, however, with considerably bigger blocks or boulders. Surely rockfalls are not everyday events in high mountains, but more frequent than suspected nonetheless. At many walks in the mountains one can notice the boulders of relatively big rockfalls (sometimes as big as a house) beside the way. Occasionally the rockfall boulders reach the surface of glaciers and then form a rough blockish

Fig. 4.12 Karst cave. The most impressive and still most mysterious results of limestone solution are the karst caves. At first rain water develops its solution activity due to its carbon dioxide content at the rock surface and the superficial karst scenery arises. After that the lime-saturated water reaches available joints and clefts in the rock and flows towards the karst water level more or less vertically. This zone between the ground surface and the water table is described as the vadose zone (of vadosus Latin. = shallow). This is actually a rather misleading name, since it can easily involve ranges greater than 1,000 height-meters, as is the case today in the Northern Limestone Alps. When the water reaches the karst water table in the mountains, solution activity is amplified. This is possible, because of the mixture between the percolating waters and the groundwater which have different temperatures and CO_2 contents. The result is a renewed solution ability, called mixture corrosion. The speleologist (of Greek. spélaion = cave) describes this as a phreatic cavity solution (of Greek. phréar = fountain) and an underground network of cavities will be formed. Since the joints and cavities are (largely) filled with water, the solution takes place in every direction. The cavities develop gradually into cave-sized depressions. Accompanied by tectonic uplift in the area, graded karst water bodies which have respectively direct reference to the previously higher water table can develop under each other. For the caves formed in the formerly phreatic zone this means that they now are in the vadose zone and are used by the percolating water into the depth as conveyers to the current (means new and deeper!) karst water level. Therefore, cave brooks or cave rivers can develop whose beds are deepened by erosion (Latin. erodere = to gnaw off). So different cave horizons have to be brought in harmony with tectonic uplift phases and the landscape development; the oldest is the most highly situated cave horizon. The photograph shows the "Verzweigungshalle" in the Salzgrabenhöhle of the Berchtesgaden Alps, Germany. It is typical of cavity formation in the phreatic zone by mixture corrosion; here the elliptical room section is formed at the crossing of formerly aquiferous clefts and a water-filled layer joint at the water table. Published with kind permission of © Dr. Ewald Langenscheidt 2014. All Rights Reserved

upper moraine. They are transported elsewhere by the glacier from their point of origin and matured by it far away again. The result is a rockfall moraine (Figs. 4.19 and 4.20).

Rockfalls can attain really large dimensions. A picturesque mountain scene can be altered completely and converted to a rocky desert within few minutes. And, in many cases, the rocks do not fall but slide at more or less steeply inclined pathways to the valley. In principle one therefore distinguishes between a fall and the much more frequent slide.

In case of fall, the rock masses immediately get with speed of fall into movement. The slide starts with a sliding movement at a path previously sketched out. The rockfall mass can remain as one or disintegrate completely on its downward journey. Frequently, bigger rock packages stay together on less steeply inclined pathways. The air included in the fall can reduce the friction between particles within the rock masses so much that they begin to flow dryly.

On both sides of the pathways of rockfall masses, embankments with a form like a moraine edge can occur. The spatial extent (or reach) of the rockfall masses increases for steeper and longer pathways and also for larger masses of falling rock. Smashed rock masses, as described above, can even reach the flat valley ground. Many rockfall avalanches were caused by melting glaciers at the end of the last ice age. The glaciers lowered the valley ground and steepened the valley sides, resulting in a U-shaped valley. When the ice melted, the walls lost their counter-bearing and became locally unstable and rockfall avalanches resulted. The deposition area of rockfall avalanches often takes the form of a

Fig. 4.13 Oxidation weathering. Gruppo di Fanis (Fanis Gruppe), Italy. The color difference between fresh rock surfaces and weathered surfaces, which becomes evident after a rock fall or within a quarry, indicates the influence of chemical weathering by oxidation. Iron is one of the most frequent elements in rocks. The brown, red, and yellow colors of ferrous oxides are also prevailing in soils. Published with kind permission of © Dr. Alexander Stahr 2014. All Rights Reserved

Fig. 4.14 Chemical biological weathering. The root secretions (the exudate) attack the rock. Published with kind permission of © Dr. Alexander Stahr 2014. All Rights Reserved

small hilly landscape with big rock boulders of different sizes. Frequently rivers are dammed by rockfall masses and lakes arise.

Big waves are a frequently concomitant with rockfalls, if the masses glide into deep waters. Rock masses repeatedly fell into Lake Lucerne (Switzerland) and caused great damage during the eighteenth and nineteenth centuries. A well-known example is the rock slide into the Italian Vaiont reservoir in 1963. Three hundred million cubic meter rock slipped into the lake, which splashed over the intact barrage after that. The wave destroyed the village of Longarone and several others and 3,000 inhabitants were killed. Norway is still more deeply affected by such catastrophes with its numerous steeply inclined fjord walls. In 1936 one million m^3 of rock brought down off the Ramnefjell slid into the west Norwegian Lovatn Fjord, for example. The rockfall caused a $\approx$70 m high wave which destroyed several settlements and farms and 74 people died. In August 1987, a violent rockfall happened at the 3,066 m high Pizzo Coppetto in the Italian Alps which devastated wide areas of the Val Tellina in Lombardia, Italy (Figs. 4.21, 4.22, and 4.23).

4.2.2 Slides

Landslides are a frequent form of mass erosion in high mountains. They are a form of mass self-movement just like falls. Therefore, they involve the removal of unstable rock or soil masses from the slope caused only by gravity

Fig. 4.15 Rockfall gutters. Dolomites, Italy. Rockfall happens frequently in so-called rubble or rockfall gutters so that debris cones are formed below these gutters. Rockfall gutters are often bounded at tectonic lineaments, such as faults or strikes. The night frost leads to frost shattering of the rock. As long as the frost remains, however, the rock is still held together by the ice. With the beginning of a new day and its associated solar radiation the temperature rises and the ice melts. Therefore, eastern and southern walls are particularly prone to rockfall in the early morning hours. Rain, storm, and thunderstrike can increase particularly the likelihood of rockfall in areas with rockfall gutters. Published with kind permission of © Dr. Alexander Stahr 2014. All Rights Reserved

Fig. 4.16 Rubble slope. Geislerspitzen, Dolomites, Italy. The rock fragments broken out by physical weathering cannot remain in outstanding walls. They are brought down as rockfall and form rubble tips at the wall's foot. Depending on the rock roughness, the surface of a debris cone can have an inclination between 25° and 40° to the horizontal. If several adjacent rubble tips grow together at the wall foot, a so-called rubble margin will be formed. Published with kind permission of © Dr. Alexander Stahr 2014. All Rights Reserved

Fig. 4.17 Sorting of material. Dolomites, Italy. Rock fragments removed by weathering and carried by gravity down to the foot of a rock face form a rubble slope over the course of time. This consists of a range of grain sizes which vary from coarse blocks and stones to sand and the finest rock particles. During the fall the coarser components are deposited furthest downslope because they have a larger mass and consequently a greater kinetic energy. The fine components accumulate in the upper area of the rubble slope. Published with kind permission of © Dr. Alexander Stahr 2014. All Rights Reserved

Fig. 4.18 Overgrown rubble margin. Dolomites, Italy. Due to relatively slow growth of rubble tips or debris cones the vegetation can more or less intensively make a foothold. Only specialized plants are able to establish themselves on the rubble tips. All of them stand out due to a high regeneration ability. Some of these plants are *Thlaspi cepaeifolium*, *Arabis alpina*, *Hieracium intybaceum*, *Dryas octopetala*, *Saxifraga oppositifolia*, and *Ranunculus glacialis*. Published with kind permission of © Dr. Alexander Stahr 2014. All Rights Reserved

Fig. 4.19 Rockfall boulders. Appenzeller Alpen, Switzerland. During the uplift of different rock layers in the course of mountain formation, variably strong tensions occur within the layers. Clefts and joints in the rock tear up and offer weathering to do its work. Comparable structures involved in the beginning of the weathering processes can have been caused by an environmental change during deposition of the sediment. If the inner cohesion of the rock body is lost a rockfall can result. Published with kind permission of © Dr. Alexander Stahr 2014. All Rights Reserved

Fig. 4.20 Breakage and rockfall. Mühlsturzhörner, Berchtesgaden Alps, Germany. In the year 1999, 200,000 m^3 of rock was brought off the peak area and crashed down to the valley. Published with kind permission of © Dr. Ewald Langenscheidt 2014. All Rights Reserved

Fig. 4.21 Rockfall landslide. A rockfall landslide in the Italian Val Tellina provided worldwide sensation in the media in August 1987. Fifty million cubic meter rock brought down from the 3,066 m high Pizzo Coppetto to the valley ground and buried several villages. The rock masses had come loose at a height of about 2,400 m up the mountain on July 28. They surged at the counter-slope another 300 m up and finally covered the valley about 3.5 km in length. The rock boulders coated occasionally with more than 100 m thickness the former valley ground. The demolition niche suffices down to 1,100 m. Twenty-eight people were killed by the rockfall. Most inhabitants however saved themselves since the area had been evacuated in time, after warnings by geologists. The steepness of the mountain flank was a cause of the enormous landslide at the Pizzo Coppetto. The catastrophe was apparently triggered by the precipitation of the rainy summer of 1987. Over many years the rock had already dropped down in centimeter steps. The long announcement for the fall cited a crevice which had already been mapped in the context of a geological thesis over 20 years previously. No expert could say, though, when it would happen. Got attentive by smaller rockfalls four days before the big rockfall landslide geologists registrated a movement of the rock masses of 1.50 m in less hours and then the alarm was given. Published with kind permission of

and without participation of a transport medium. Although running water is excluded from the process of sliding by this definition, nevertheless it plays a decisive role in the enabling of landslides. This is because it reduces the friction (or cohesion) between rock or soil particles and increases the weight of rubble or soil masses. It also exerts flow pressure on individual components of a loose material at the slope.

As a rule, the triggering factors of slides are changes in the slope balance. Over long time spans these can be caused by chemical and physical processes of weathering which led to an over-position of slopes, or in the short term by precipitations, earthquakes, and human influence (e.g., by road constructions). As long as the counteracting forces are greater than the driving forces for a landslide, a slope remains stable. If the driving forces become greater, a slide happens. At this moment shearing between two rock or soil layers occurs and clods glide relatively fast to the valley ground.

Essentially there are two forces which counteract a sliding process: cohesion and inner friction. Both prevent the slipping of rock or soil material and are therefore responsible for shear strength, which means the resistance of the material to moving. Cohesion can be explained by two causes. On the one hand there is the electrostatic force between the clay minerals. On the other hand, there are water menisci between the fine particles (up to sand grain size) which one also describes as an apparent cohesion.

The friction is influenced by the form of the rock and soil particles and their respective sizes. The more rough and/or edged components are in the mixture, the bigger is the inner friction of loose material. If a slope only consists of rough rubble, the friction works contrary to movement. There are even broader factors in addition to cohesion and friction which increase shear strength of loose material at the slope: Tree roots for example hold soil or rubble masses together as well. Humus particles and minerals which are dissolved in the groundwater and precipitated upon drying out can stick the loose fine rock or soil grains to each other. Together all these factors constitute the sum of forces counteracting movements.

The weight of material and the slope inclination support a slipping process. If it rains heavily, water penetrates into the loose masses of a slope and changes the inside balance. Depending on how much loose material a slope consists and how much water can penetrate into it, the shear force will be reduced considerably. The apparent cohesion disappears at large water quantities and the electrostatic attraction between the clay particles is lowered. Water

Fig. 4.22 Deposition area. The photo shows the deposition area of the rockfall landslide at the Pizzo Coppetto, Val Tellina. Nothing more can be recognized of the original "green" valley ground. The boulders of the rockfall fragmented to rubble during the slip and now cover the valley along a 3.5 km stretch. The rubble masses in the valley caused an additional danger because they dammed the river Adda. From the lower Val Tellina over 20,000 inhabitants were evacuated since they were threatened by the break of the dam. Because of intense rain, the water level of the new lake grew faster than expected. However, by a spectacular action the threatening catastrophe was prevented. A controlled flood through the crown of the dam allowed the torrents of water to drain at a safe speed. Excavators sank a break into the rockfall masses before the lake was filled, allowing the lake to overflow and drain away in the river bed of the Adda. Published with kind permission of

Fig. 4.23 Rockfall lake. Lake Pragser Wildsee, Dolomites, Italy. The "Pragser Wildsee" is located 1,496 m above sea level in the South Tyrolean Dolomites. One must look at this lake more closely to recognize, as with many other romantic high mountains seas, the catastrophe which led to their emergence. A rockfall formed the lake whose boulders can still be recognized at its northern end. Rockfall lakes are among the most frequent results of rockfall landslides. Published with kind permission of

Fig. 4.24 Translation slide. Pustertal, Osttirol, Austria. Translation slides are mass movements on a preexisting slip plane. The gliding area can be the boundary between a surface layer of loose material and the basement rock (with different grain sizes), the lower level of plant roots, or the transition between two rubble layers. Also the boundary between more permeable and thick stored layers can appear as a potential slip plane. This form of slide is also possible in solid rocks, if between two rock layers an interlayer is stored, for example, clay. Published with kind permission of © Dr. Alexander Stahr 2014. All Rights Reserved

Fig. 4.25 Rotational slide. Berchtesgadener Land, Germany. The gliding areas of rotational slides are roughly circular (in cross section) unlike those of translation slides. They arise only at the moment of slip since mechanically effective boundary layers are absent within the loose rock. The slide mass undergoes a rotation around a slope parallel axis, which leads to the tilting of the material along a concave line. Published with kind permission of © Dr. Alexander Stahr 2014. All Rights Reserved

flowing downwards in the slope puts pressure on each of the individual components simultaneously. This so-called flow pressure, together with the weight of the loose masses and the inclination of the slope, causes a net downslope force. Once stronger than the restraining forces, the slope starts moving. Disturbances caused by earthquakes, volcanic eruptions, or building projects can also adjust the balance of forces in a slope decisively and trigger slides.

In principle two types of slides can be distinguished: translation slides and rotational slides. In case of translation slides one layer slips over another. The gliding area is already therefore preformed (a preexisting feature). For example, this can be the boundary layer between gravels permeable to water and a bed of sodden clay. As a general rule, the gliding areas accommodating translation slides are aligned roughly parallel to the surface of the slope.

Slides behave differently at slips in homogeneous material. Here the gliding areas are more or less circular arcs. One describes this type of slide as a rotational slide. The gliding areas of rotational slides arise at the moment of slippage because mechanically effective preexisting layer boundaries are absent in the loose rock. The slide mass undergoes a rotation around a slope parallel axis, following a concave path so that the slipping material becomes tilted.

To counter slips, man has devised measures and methods to keep endangered slopes stable. They range from careful dealing with the protection of forest zones up to the biologically engineered obstruction of dangerous slopes. Japan is surely one of the leading nations in the field of technical slope stabilizing. With its mountainous islands, unconsolidated volcanic sediments, and earthquakes, the country is often struck by disastrous landslides (Figs. 4.24 and 4.25).

4.2.3 Flowages

In nature the classification of different slope movements is frequently very difficult with transitional forms existing between individual movement patterns. Slides can change into flow motions or can be the result of creeping movements

Fig. 4.26 Creeping of soil. Matreier Becken, Osttirol, Austria. Very damp ground upon clay or loamy substrata can creep along the slope under influence of gravity down towards the valley, primarily, if they, how frequently given in high mountains, are very shallow. Then they quickly absorb water and become plastic. The creeping soil is often compressed into waves and folds. It frequently gets rips which are more or less big and situated crossways to the slope. Vegetation appears, prefering damp to wet ground. As a rule, creeping movements take place very slowly till extremely slowly. Since they depend on the water supply, they take place occasionally even only episodically. This is recognizable at the so-called saber growth of trees if creeping of soil takes place under woods or under solitary trees. By the slow, partly episodic movement of the ground, the trees have sufficient opportunity to grow always in the direction of light. In the course of years the striking saber growth arises, like it is recognizable in the photo. However, saber growth or slanted trees can be watched also at other slow slope movements like the movement of whole valley flanks, so that a more thorough examination of the area gives first references to the way and exact cause of the movements. Published with kind permission of

of the ground. Geoscientists, forest scientists, and engineers from all over the world have dealt already for many decades with the problem of classification of slope movements, so far without coming to a completely clear distinction between the movement patterns. As with the classification of volcanoes, nature can be pressed only with great difficulty into a scheme.

Besides the speed of motion, a distinctive feature of mass shifts is their movement mechanics. In this manner, slides are in principle different from motions of *flow* which take place largely without brittle deformation (cracking and crumbling). Here the rock or soil material is more subject to a plastic or essentially viscous movement. Depending on their water content, rubble or soil masses are able to move downwards like tough mash or thin honey. Motions of flow often take place very slowly, over many years or decades. One then speaks about creeping movements down a slope (Fig. 4.26).

Slope movements and slides are narrowly and open frequently restricted near surface layers. However, whole mountain flanks can hardly visibly glide slowly to valley ground over years and centuries. Clefts and rifts as a note of the movements gape in the mountain from time to time. One describes such big movements as movements of valley flanks ("Talzuschub" in German). Mountain flanks can slowly glide in ranges of several 100 ha to the valley. These extensive and from time to time more than 100 m in the depth reaching movements were triggered by melting glaciers at the end of the ice age. The glaciers over-steep the valley flanks and formed trough valleys. The counterbearing was missing after the retreat of the glaciers. It came to rockfalls, landslides, and slow slope movements at the over-steep valley flanks. While the characteristic U-shaped form of the glacially valleys was remaining unchanged in stable rocks, creeping movements started in strongly tectonically ruined and in softer rocks, which partly last until today. Also the destruction of mountain flanks by rivers at their banks can trigger these big landslides. Where rock masses on clayey layers are creeping down slowly, slides can occur in case of strong rainfalls out of the slow movement. All slow motions of creeping were indicated by relatively narrow surface areas.

An also narrow surface area with numerous flowing bulges causes the solifluxion (of Latin solum = soil and fluere = to flow). This is a slowly creep because of seasonal or daily frost change. Therefore, this form of mass movement is restricted largely to the periglacial region which already starts at an altitude of sea level in arctic high

Fig. 4.27 Solifluxion. Furkapass, Switzerland. Slow flow of water-saturated loose material on a still frozen basement is described as solifluxion. Solifluxion therefore takes place in high mountains in the periglacial region, at an altitude which is characterized by frequent frost changes. While these frost changes stop largely in the middle latitudes and in the Arctic during winter, they are effective in the tropical high mountains daily. The active saturated layer of the free solifluxion forms a flowing earth with striking solifluction lobes. Published with kind permission of © Prof. Dr. Heinz Veit 2014. All Rights Reserved

mountains. By the melting of the frozen soil the water-filled layer of the soil surface gets into flowing. Striking solifluxion sutures, real flowing tongues, and solifluxion scree are formed. There are free and bound solifluxion. The motion of flow is made easier at free solifluxion since vegetation is missing. Bound solifluxion takes place over the timber line under grass and bush vegetation. The flowing earths cause often the tearing up of turf and the formation of striking flowing bulges. Besides flowing plastically the frost piston stroke also plays a not insignificant role for solifluxion. Soil components are raised vertically during the freezing process to the slope surface. When melting they sink back following vertically the gravity, so that they have moved of their original place the slope downwards (Fig. 4.27).

To the fast flow motions belong debris flows and mudflows. As mud flowing the fast shift of water filled up loose masses of the slope is understood, what happens in course of heavy rainfall. Mudflows frequently arise from clayey marly rocks. They happen in the Flysch of the northern Alps, for example. The speed of mudflows varies from some 10 m to more than 100 m a day. Initiators of these movements are often extensive rotational slides in profoundly weathered stratas of clayey marly sediments (Fig. 4.28).

Scree flow is not restricted exclusively to a slope but also occurs in thick infillings of rough clastic material (Greek: klásis = to break), for example, the Wimbachgries between the mountains Watzmann and Hochkalter in the Berchtesgaden Alps (Germany). The scree flow has a length of over 10 km and in places attains a thickness of more than 300 m. It consists of polygonal rock fragments of Ramsau Dolomite, coming from both valley flanks by gravity and mechanically weathered there by frost blasting. During heavy rainfall parts of the scree flow are remobilised again, the scree really swims up. Approximately 4,500 tons of scree leave the valley every year through a gorge (Figs. 4.29 and 4.30).

Fig. 4.28 Debris flow. The photo shows a debris flow near the upper Bavarian place Inzell. 50,000 m^3 of thick rubble moved at speeds of up to 50 m per day downwards from the 1,265 m high Teisenberg at the beginning of August 1991. Triggers were heavy rainfalls in the Bavarian Alps with 100–200 mm precipitation between July 30 and August 3, 1991. By the torrents of water the profoundly weathered sandstones and marls of the Flysch zone at the west flank of the Teisenberges get intensively wet. The debris flow moved at about 40 m a day towards the settlement Hutterer. The flow finally stopped in the middle of August, because the water running out a source could be drained. The starting point of the flow was a historically active large slide in the weathered Flysch sediments. Published with kind permission of © Dr. Ewald Langenscheidt 2014. All Rights Reserved

Mudflows are also extremely rapid flows which appear after strong rain or with the melting of snow. A mudflow is a mushy mixture of water, soil, rubble, larger rocks, and other materials dragged along (such as bushes and tree trunks). They contain more water than solid material. It therefore is a suspension (Latin suspensus = floating) — the aggregates float in the water.

Fig. 4.29 Scree flow at the Wimbachtal Valley. Berchtesgaden Alps, Germany. Extremely rapid movements of thin, mushy, water-oversaturated loose masses of steep slopes are also described. They were triggered by extremely heavy precipitations. Normally the initial stage is a slide. The primary slide masses push away the slope downwards lying soil which immediately flows away because of the high water content and porewater pressure. The slipped material flows like a mudflow, however, not in gutters and grooves but free along the slope. Published with kind permission of © Dr. Ewald Langenscheidt 2014. All Rights Reserved

A mudflow derives its material mainly from catchment areas with little vegetation. A rubble dam is frequently deposited on either side of the path of a mudflow since the flow moves considerably more slowly at its edges. Mudflows running outwards can form a fan, and volcanic mudflows, called lahars, are comparable to fans.

Mudflows often follow the course of slope grooves or wild river beds. Therefore they are more or less channeled. At bottlenecks in the valley course mudflows can be caught and dangerous dams can result. The dam brakes when the enormous pressure of the suspended masses and torrents of water lying behind it become too large. At the breakthrough point, mudflows can reach speeds up to 100 km/h and move boulders with a weight of 100 tons. Disastrous devastations are the consequence (Figs. 4.31 and 4.32).

4.2.4 Abrasion and Transportation

Abrasion causes loosening of rocks. Through transportation the removed rocks are moved away. In nature both processes are interlinked and difficult to separate from each other.

Fig. 4.30 Mass shifting. Mass shifting in the Tyrolean Alps (Matreier Becken). Such extremely fast mass shiftings do seldom take place under closed forests (Waldabbrüche). The shifted masses can be recognized in the picture center on the wooded slope of an old valley. The demolition area gets more broad downwards — here it is about 2 ha wide. Published with kind permission of © Dr. Alexander Stahr 2014. All Rights Reserved

An essential process of abrasion and transportation is in many ways soil erosion. Soils with a high silt content are particularly vulnerable to erosion by water. Clayey soils show a greater cohesion; soils with sand and gravel have a higher grain weight and a better infiltration rate so that in the two cases the erosion is aggravated by water opposite the silty soils. Soils made of loam which have an even distribution of all grain sizes are less susceptible to soil erosion by water. Layer erosion is a gradual two-dimensional abrasion by the combination of raindrop impact (splash erosion) and surface runoff. Groove erosion is made by running water in micro deepenings which expand at continuously running water. In extreme case real erosion ravines arise, the drains. Soil erosion frequently takes place in high mountains on areas which have been robbed off their vegetation before by other mass shifts such as landslides, avalanches, or snow abrasion. The condensed, partly vegetationless soils of pathways, meadows, and ski runs are as well affected by running waters (Figs. 4.33, 4.34, and 4.35).

Earth pyramids are a striking result of runoff processes. They arise from erosion of thick moraine deposits of the ice ages. If rougher moraine blocks are exposed, they temporarily protect the material lying underneath from erosion so that impressive pyramid-shaped buildups result (Fig. 4.36).

Erosion by flowing waters leads to the deepening of the bottom and simultaneously at the sides. Through this it can come to the brake down of embankments robbed of its stability (Fig. 4.37).

Another form of erosion, often seen on alpine meadows, are the cattle trails. They are the result of constant kick load to the ground due to animals which graze parallel to the slope. During damp weather cows can kick off bigger clumps of soil together with vegetation and this can be the starting point for other forms of erosion such as snow abrasion. Eventually the interplay of these processes leads to the formation of irregularly vegetationless areas on the slope (Fig. 4.38).

Fig. 4.31 Mudflow. Mudflows are extremely rapid motions which occur during heavy rainfall. They constitute a mushy mixture of water, soil, rubble, rock chunks, bushes, and trees dragged along in suspension. The water content of a mudflow exceeds its total solid content. The photo shows the deposits of a mudflow in the Berchtesgaden Alps. Published with kind permission of © Dr. Ewald Langenscheidt 2014. All Rights Reserved

Fig. 4.32 Lahar. Mount St. Helens, Cascade Range, USA. Similar in appearance are volcanic mudflows. They can reach speeds of more than 50 km/h. During heavy rain on steep slopes lahars arise from volcanic ash or by melting ice at a volcanic eruption. Published with kind permission of © U.S. Geological Survey (USGS, public domain) 2014. All Rights Reserved

Fig. 4.33 Erosion on a so-called Blaike. Berchtesgaden Alps, Germany. The photo shows two different forms of erosion. Besides several runoff gutters which arose from heavy rainfalls in the summer months, the "Blaike" (a German word) shows a stone plaster as a result of losses of fine material due to layer erosion. The "Blaike" itself arose from a snow slide during the thaw in spring. Published with kind permission of © Dr. Alexander Stahr 2014. All Rights Reserved

Fig. 4.34 Soil erosion on pathways. The soils and the ground of pathways are strongly compacted. By standing or going the mechanical pressure load of the ground reaches punctual values of up to 400 g/cm^2. Even amounts of up to 57,000 g/cm^2 were obtained by increased dynamics of the individual kick at the way downwards. Rain and snow water cannot seep away any more and drains away superficially. Little soil plates are broken off and taken away at the pathway surface. Deep erosion grooves which can be the initial stadiums of further erosion processes are often the consequence. Silty soils primarily are endangered by soil erosion. Published with kind permission of © Dr. Alexander Stahr 2014. All Rights Reserved

Fig. 4.35 Trenches. Marienschlucht, Allgäu, Germany. Moraines, gravels, and slope rubble often build up steep and high valley flanks. Mass movements like slides frequently cause primary damages to soil and vegetation cover in these areas. If thick loose rock masses are exposed once, erosion by water can become effective, up to extreme extents from time to time. Precipitation cuts quickly deep erosion gutters into the vegetation-free areas. Because of erosion single gutters can unify to form deep cut trenches and erosion ravine units. Motions of flow and rockfall also can be involved in the further erosion of loose material. Published with kind permission of © Dr. Alexander Stahr 2014. All Rights Reserved

Fig. 4.36 Earth pyramids. Nearby Bolzano, Southern Tyrol, Italy. Thick moraine deposits are subject to the process of downwash at a missing vegetation cover. The material of the moraines consists of unsorted rock fragments which are composed partly of larger stones or blocks. If such a block is exposed by erosion, it temporarily protects the material underneath from further erosion. The striking earth pyramids were formed this way over course of time. If a pyramid becomes too high, its slopes too steep, or its base too unstable, the whole structure falls down. Published with kind permission of © Dr. Alexander Stahr 2014. All Rights Reserved

Fig. 4.37 Erosion of a bank. Allgäu, Germany. Shore demolitions are the result of erosion by flowing waters. On the upper edge it causes slipping of loose material; however also the sub-flush of groundwater leads here to overturning trees. Published with kind permission of

Fig. 4.38 Erosion by the offset of clods. Northern Limestone Alps, Germany. The offset of clods together with vegetation is caused by the kick of man and grazing animals. This forms the starting point for further erosion processes. The thin soil is then taken down and vegetationless rubble areas stay behind. This is favored by continual pasturing of areas which are already damaged. Published with kind permission of

5 Landscapes Caused by Geological Structure and Rock Features

A. Stahr and E. Langenscheidt, *Landforms of High Mountains*,
DOI 10.1007/978-3-642-53715-8_5, © Springer-Verlag Berlin Heidelberg 2015

Geological structures and different geomorphological features of rocks imply a clear differentiation of predominately climate-controlled weathering and erosion processes. This applies not only to high mountains, but also to all other terrestrial landscapes.

Weather-resistant rocks are eroded slowly and to a lesser extent than those more susceptible to weathering and erosion. Lithological diversity within a mountain range can cause great variation in the appearance of peaks from one region to another.

In one mountain region, steep faced rugged peaks can coexist with softer landforms that resemble submountainous areas. In the same mountain area we can see on the one hand softer and smoother landforms, rather like low mountain ranges while nearby rugged and steep-sided mountains grow into the sky. The imposing towers of the Dolomites for example, with their occasionally overhanging walls, present great contrast in comparison to the softer shapes of the crystalline Central Alps. Furthermore the geological structure is another key influence in shaping the landscape. It dictates the path that valleys follow as well as the shape of high mountain peaks. The development of mountain landscapes is thus the product of weathering and erosion on structurally and lithologically inhomogeneous rock bodies.

5.1 Structure-Controlled Development

Geological structure at upper crustal levels is dictated by the orientation of different rock bodies and the tectonic forces to which they are subject. Separation planes or fracture systems created in the course of mountain building can have substantial geomorphological influence. As illustrated in Chap. 2, geomorphological features are often the manifestation of the internal structures by which they are controlled. Such structures include folds, fracture systems or the stratification of a mountain. These tectonic structures create conditions that promote weathering and erosion.

Let us turn once more in this respect to the Dolomites. The rock walls of this mountain range, which often arise as bizarre rock crags and towers, are due to jointing. The uplift of the Dolomites started in the late Cretaceous period (some 70 million years ago) and took place in different phases. Uplift led to severe jointing of the carbonate rocks, allowing freestanding towers, smooth rock walls, or a variety of sharp rock spires to be weathered out.

Fissures, fractures, faults, and large-scale tectonic lineaments compromise lithological coherency. Such structures enable the effects of erosion to reach great depths. Therefore, the course of valleys, passes, gorges, and river networks is also often closely associated with these structures. The straight-lined valley Gailtal, nearly 110 km long, follows in the Eastern Alps the deep-reaching fault of the Periadriatic Fault (western part known as Lesachtal).

Geological structure dictates the drainage network of river systems. Right-angled faulting or fracture systems influence the incision of a river to an extent that its path is characterized by straight lines and sharp bends. This type of structurally influenced river network is referred to as a rectangular or trellis drainage network. The path of dendritic drainage systems is equally structurally influenced, as is that of radial networks, which follow the strike of underlying strata.

Strike

The term "strike" (also: stratum contour) is used together with the term "dip" to define the orientation of geological structures such as bedding or cleavage in space. Strike is the intersection of a planar surface with an imaginary horizontal plane. Dip defines the inclination of such structures and is perpendicular to the strike.

Like water, glacier ice follows the tectonical structures. The path of glaciers is even more prone to the influence of geological structure. Therefore, the glacially ravaged landscapes of Scandinavia, Greenland, New Zealand, and Patagonia provide insight concerning the tectonic situation at lower levels. Such structure-controlled landscape formation is manifested in the form of the famous fjords.

Thick sedimentary packages of many mountain ranges consist of a vertical sequence of alternating beds due to changes in depositional environment. The alternation of softer and harder rock layers leads to development of steps or ridges; softer strata are eroded, leaving harder ones to create terraces. In this case, landscape forms are controlled by stratification. Morphologically softer rocks are poorly permeable sediments such as shales. Fractured sandstones or limestones on the other hand are water permeable and thus provide significantly more resistance to erosion. They are therefore morphologically hard rocks. If flat lying strata of different morphological hardness are subject to fluvial or glacial erosion, differential weathering and erosion, corresponding to varying lithology, commence at the exposed flanks. The softer layers are more affected by erosion; over time, their removal leads to the denudation (from Latin denudare = to be bare) of the harder strata more resistant to erosion. The result is a stepped profile with a terraced slope. Such platforms are called denudation terraces.

Sedimentary strata, once deposited flat on the seafloor, can be folded, metamorphically altered, and thrust over other nappes during the course of orogenesis. With the onset of

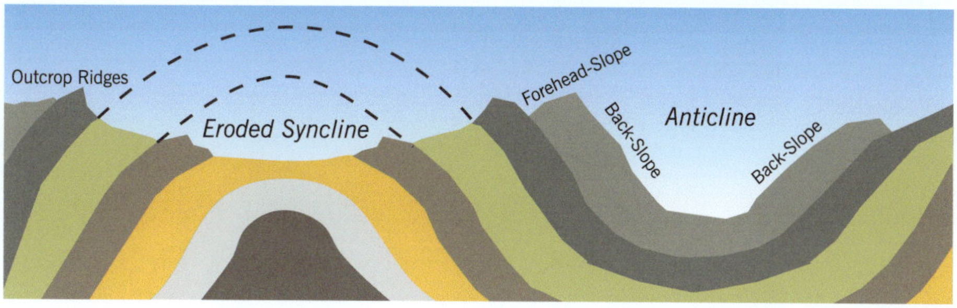

Fig. 5.1 Reversal relief structures and hogbacks (schematically). Published with kind permission of © Dr. Ewald Langenscheidt 2014. All Rights Reserved

uplift to the mountains, these structures come to surface, where differential erosion removes less resistant rocks and denudation begins. Especially in highly stressed structural areas, such as fold saddles or nappe faces, weathering and erosion are extreme. The result is often a striking geomorphological phenomenon, the so-called inverted relief. Inverted relief is reflected by the fact that landscape and tectonic structure do not match. The erosion of the saddle of an anticlinal structure can result in a valley that follows the strike of the antiform.

The antiform can often only be recognized when the layers exposed at the valley walls are used for reconstruction. The process of relief inversion exposes strata boundaries at the more or less steep standing limbs of antiformal structures. Erosion of a tilted or vertical series of alternating soft and resistant beds leads to the formation of ridge forms such as cuestas and hogbacks (Fig. 5.1).

Ideally, hogbacks are found where the beds are nearly vertical. In this case, the dip of each flank of the hogback is identical and a steep symmetrical ridge is the result. It is much more frequent however that the limbs of an antiform have contrasting dips. A cuesta is an asymmetrical ridge, its gentle slope is determined by the dip of the resistant bed, and the opposite side is a steep escarpment formed at the erosional edge of the resistant bed where it is undercut by erosion of the softer bed beneath.

In large-scale thrust systems and at nappe fronts, even when no considerable folding is present, hogbacks are the result of tilted beds. There is a tectonically controlled asymmetry of peaks, a prime example presented by the dolomites.

Between the forms of denudation terraces with their horizontal sequences and hogbacks with vertical layers, all transitions are possible. Therefore, all transitional forms with inclination of the rear slope greater than 5° are referred to homoclinal ridges (from the Greek homos = same, together and klino = tend). Beds that dip with less than 5°, as it is often the case in large-scale crustal flexures, are referred to as escarpment landscapes. In contrast to submountainous areas, escarpment landscapes play a subordinate role in high mountain regions. In such regions they either are merely a local phenomenon or transition to hogbacks near complex fold structures (Figs. 5.2, 5.3, 5.4, 5.5, 5.6, 5.7, 5.8, 5.9, and 5.10).

5.2 Shaping Due to Rock Features

In the previous chapter, the influence of lithology on a rock's resistance to weathering and erosion was briefly introduced. When approaching the Bavarian Alps from north, gentle undulating submountainous landforms dominate, whereas further south, rugged steep-walled limestone peaks rise into the sky. The undulating submountainous areas owe their shape to the Flysch by which they are built and are commonly densely forested. The mixture of marls, shales, and sandstones intercalated with clay makes the flysch sediments vulnerable to weathering and erosion. Constant landslides prevent the formation of steep slopes.

Further south in the Northern Limestone Alps there are two main landscapes, whose habitus is again directly coupled with the lithologies of the rocks found there. Triassic limestones (Wetterstein Limestone, Dachstein Limestone, etc.) form the main mountains and steep faces and, due to their purity, have been subject to karstification. In contrast, the local Jurassic sedimentary rocks (marls, cherty limestones, etc.) weather to clay-rich loamy substrates. The deep soil cover created by these siliceous limestones leads to softer landforms than true limestone does. Because of this, but also due to the wealth in springs and perennial surface waters, these areas are most commonly used as alpine pastures. In many areas of the Northern Limestone Alps, patterns of anthropogenous utilization reflect the geological circumstances, particularly the distribution of forests and pastures. Sediments prone to weathering such as marl, siltstone, or shale create similar landforms in the Dolomites as well, providing great contrast to the steep carbonate walls. Similar contrasts are found in the Central Alps. Morphologically hard prasinites (metamorphosed ocean floor basalts) of the Glockner Group build steep peaks with sharp, sometimes

Fig. 5.2 Gorge. Aschauer Klamm, Berchtesgaden Alps, Germany. A gorge is a type of valley with a steep V-shaped cross section. The valley walls are vertical and so close that the river completely fills the valley floor. Such valley shapes are usually coupled with zones of tectonic weakness. Fluvial incision is relatively fast and is too young to have flattened the walls of the gorge. A further condition for canyon or gorge formation is the sufficient stability of rocks that are not prone to landslides. Published with kind permission of © Dr. Ewald Langenscheidt 2014. All Rights Reserved

Fig. 5.3 Rock towers. Fanes Group, Dolomites, Italy. The carbonate rocks of the Dolomites have been riddled with innumerable crevices during tectonic uplift. Weathering and erosion took place at these fissures and ultimately created freestanding rock towers with vertical walls and a variety of rugged stacks. Published with kind permission of © Dr. Alexander Stahr 2014. All Rights Reserved

jagged ridges such as the Großglockner itself. Other high peaks in the Central Alps are composed of resistant rocks such as gneiss or granite. Prominent examples are the Venediger Group, the Granatspitz Group, or the Rieserferner Group. In contrast to the prasinites, easily weatherable mica schists form very rocky, steep slopes of up to 40° inclination that bear little or no vegetation where intensive frost weathering occurs in connection with eolian erosion. In the central alpine Ortler Group, morphologically hard metamorphic carbonate rocks form precipitous faces and imposing massifs (Ortler, Monte Zebru, Königsspitze), whereas the morphologically softer quartz phyllite creates undulating shapes.

The physicochemical lithological circumstances also determine the extent of jointing. A prime example can be found in the Berchtesgaden Alps, at the 2,713 m high Mount Watzmann. The Ramsau Dolomite forms its base, followed by the Raibler Beds, which are overlain by Carnian-norian Dolomite and finally the wall-forming Dachstein Limestone. In contrast to the Dachstein

Fig. 5.4 Alpine valley. Dolomites, Italy. Fractures, fissures, faults, and large-scale tectonic fault lines weaken the rock, providing weathering and erosion an easy path of attack. Hence, the path of great longitudinal valleys of the Alps and other mountains is determined by large-scale geological structures such as nappe boundaries. Published with kind permission of

Fig. 5.5 Glaciers and cracks. Much finer than water feels the ice of the glaciers for geological structures such as cracks. Rocks frozen in at the base of the glacier are fractured by detraction (from Latin: detrahere = to remove) and levered out by exaration (from Latin: exarare = to furrow). It is obvious that the rocks in especially tectonically stressed areas are greatly affected by glacial erosion. The famous fjords in Norway are the result of such glacial structure shaping. Published with kind permission of

Limestone, the extensively jointed Ramsau Dolomite is more prone to weathering, resulting in increased production of talus material. This striking difference is due to the fact that the brittle Ramsau Dolomite formed fracture joints in response to deformation, whereas the softer Dachstein Limestone formed folds (opening picture of this chapter, Figs. 5.11, 5.12, 5.13, 5.14, 5.15, 5.16, 5.17, 5.18, and 5.19).

Fig. 5.6 Denudation terraces. Sella, Dolomites, Italy. Landscape forms of high mountains are determined among other things by stratification. If a flat lying, alternating series of soft and resistant beds are cut by erosion, erosion will commence on the exposed flanks. The more resistant layers develop steep flanks compared to the softer beds, which are subject to amplified areal erosion and denudation. Morphologically softer rocks are, for example, barely permeable sediments such as marl and claystone. Morphologically hard rocks, like fractured sandstone or limestone however, are water permeable and thus provide significantly more resistance to erosion. The result is a gradual steplike profile, softer beds are eroded, and denuding resistant ones form denudation terraces. In the case of the Sella Group (see above), the morphologically softer layers of the sandy-marly Raibl Beds were eroded creating a striking circumambulatory denudation terrace. The vertical rock walls above and below the terrace are formed by morphologically harder carbonate rocks (Haupt and Cassian Dolomite). Denudation terraces are a common landscape form in the Dolomites but also in other high mountains with alternating beds of varying hardness, for example, in the Canadian Rocky Mountains. Published with kind permission of

Fig. 5.7 Hogbacks. Venediger Group, Austria. Differential erosion of a tilted or vertical series of alternating soft and resistant beds leads to the formation of ridge forms such as cuestas and hogbacks. The hogback follows the strike of the bedding. The picture shows nearly vertical standing beds of an eroded fold structure, consisting of a series of mica schist and gneiss with different morphological hardness. The harder gneiss layers were worked out as hogbacks, whereas the softer mica schist formed the hollows. Published with kind permission of

Fig. 5.8 Structural asymmetry. Untersberg, Berchtesgaden Alps. Tectonic processes such as folding and nappe thrusting during orogenesis can transport once flat lying seafloor sediment packages to the earth's surface, often tilting the beds in the process. The morphological result is a frequently observed structural asymmetry of peaks. An example of such conditions is given at the Untersberg in the Berchtesgaden Alps (*background*). The summit of Untersberg shows a striking asymmetry with a much shallower northern flank in contrast to the steep south face. The northern flank is more or less sculpted by the bedding plane of the nappe, building the summit of Untersberg. The south face on the other hand is perpendicular to the dip of the bedding planes. Hence, the nappe forehead and the joints perpendicular to it are subject to increased weathering and erosion. For these reasons, Untersberg can be referred to morphologically as a typical layer outcrop. Published with kind permission of © Dr. Ewald Langenscheidt 2014. All Rights Reserved

Fig. 5.9 Fold structures and landscape. Magdalena Bay, Svalbard Archipelago, Norway. The different structural and geomorphological features of rocks (from the Greek gé = earth, morphé = form, and lead lógos = science, education) cause otherwise strictly climate-controlled erosion and weathering processes to have different effects on landform development. This phenomenon is manifested clearly in Svalbard. Paleozoic rocks with pronounced folding dominate the west part of the island group; the topography is rugged, characterized by countless jagged peaks. Strata structurally compromised by folding are especially vulnerable to weathering, as well as various erosional mechanisms (i.e., rock-fall or aquatic and glacial erosion. Restless and chaotic landforms are the result in such regions that are structurally compromised and hence prone to amplified weathering and erosion. In contrast, a plateau landscape formed by flat-lying sedimentary beds dominates the interior and east of Svalbard. Published with kind permission of © Herbert Funk 2014. All Rights Reserved

Fig. 5.10 Horizontal stratification and landscape. Isfjord, Svalbard, Norway. The Svalbard archipelago shows a striking duality of landscape forms, indicating different geological structures. Extensively folded rocks create rugged peaks with sharp pinnacles providing a great contrast to the plateau landscape formed by flat-lying sedimentary strata. The exposed bedding planes of the strata are for the most part structurally intact and therefore provide erosion and weathering fewer modes of attack. The alternation between resistant beds to those less resistant to weathering and erosion has led to the scree-covered slopes visible in the middle of the picture. The scree slopes indicate areas where softer beds crop out. Concentrated erosion has created small, tight V-shaped valleys of little longitudinal extension that now interrupt the once continuous strata. Published with kind permission of © Herbert Funk 2014. All Rights Reserved

Fig. 5.11 Low mountain regions within high mountains or in front of it ("Mittelgebirge"). Chiemgau, Germany. The low mountains there consist of easily weatherable rocks prone to landslides, such as marls, shales, and sandstones intercalated with clays. They are characterized by their gentle shapes. Cliffs and steep slopes are the exception. In the Northern Alps, rocks with such properties are found in the Flysch zone. The Flysch is a sedimentary sequence of erosional debris deposited in the foreland basin of a developing orogen. Repeated turbidity currents sweep the sediments from the shelf into remnant ocean basins, creating the characteristic fining upward sequences. The sequence of conglomerates, sandstones, and marls is interbedded with thin layers of clay. The clay intercalations in the coarse-grained sediments can act as detachment horizons, especially when they are water saturated and the strata is tilted. A familiar example is the debris flow at Inzell in 1992, caused by a large-scale landslide of Flysch at Mount Teisenberg after heavy rainfall. Flysch zones also occur at the edge of many other high mountains, such as the Andes or the Carpathians. Published with kind permission of © Dr. Alexander Stahr 2014. All Rights Reserved

Fig. 5.12 Wall-forming rocks. Mount Watzmann (2,713 m), Berchtesgaden Alps, Germany. In high mountains the occurrence of neighboring rocks with differing resistance to weathering creates especially striking landforms. Marl, cherty limestones, and other easily weatherable rocks create more gentle forms, whereas in the immediate vicinity in situ carbonate rocks form imposing walls. In the Northern Limestone Alps, colloquial terms were initially used to describe such striking morphological contrasts. The Schrattenkalk in Switzerland, the Wetterstein Limestone, Plattenkalk, and the Dachstein Limestone in the Bavarian and Tyrolean Alps are prominent wall-forming rocks. Published with kind permission of © Dr. Ewald Langenscheidt 2014. All Rights Reserved

Fig. 5.13 Alpine pastures. The local occurrence of marly-siliceous, easily weatherable sediments (for example, the Kössen, Raibl, or Chiemgau Beds) leads to a distinct differentiation of landscape habitus in the Northern Limestone Alps. Whilst carbonate rocks create steep to vertical walls and enable the creation of only shallow and rocky soils, marl and cherty limestones weather to deep loamy substrates, creating soft landforms. Because of these fertile soils as well as of numerous springs and mountain streams, these areals are often used as pastures. These weathered beds are therefore called "Almhorizonte," which means mountain pasture horizons. Published with kind permission of © Dr. Ewald Langenscheidt 2014. All Rights Reserved

Fig. 5.14 Shapes of the granitic Alps. In contrast to carbonate rocks, which create striking walls, many metamorphic rocks such as phyllite or mica schist form a smoother relief. The result is a strong scenic contrast between the crystalline Central Alps and the Northern Limestone Alps and the Dolomites. The granite or gneisses of the Central Alps, on the other hand, are resistant to weathering. These rocks also create steep walls; the Bregaglia poses a fine example with its famous granite peaks as well as sharp ridges and well-preserved cirques. The photo shows the 3,435 m high Hochgall in the Riesenferner Group from the west. The long ridges of granite rocks extend all the way to the valley floor. The "granites" of the Riesenferner Group are sensu strictu tonalite, a special type of granite with a different mineralogical composition which is named after its occurrence at the Tonale Pass in South Tyrol. Published with kind permission of © Dr. Alexander Stahr 2014. All Rights Reserved

Fig. 5.15 Mountain shapes in limestone. Mount Robson (3,954 m), Rocky Mountains, Canada. The summit consists of limestone and dolomite, which are separated by a reddish-yellow band of shale from the subjacent quartzite. The rocks of Mount Robson are part of a large-scale fold or syncline. Because of their location at the apex of the fold they appear to be horizontal beds. Published with kind permission of © Heiko Kurz 2014. All Rights Reserved

Fig. 5.16 Mountain shapes in metamorphic limestone. Within the Central-Alpine Ortler Group, the morphologically hard carbonate rocks, resistant to weathering and erosion, form the steep-walled imposing figure of Mount Ortler (3,905 m), Monte Zebru (3,740 m), and Königsspitze (3,859 m, left half). In contrast to the carbonates of the Northern and Southern Alps, the calcareous sediments here are slightly metamorphosed as they were heated to almost 400 °C during orogenisis and thus transformed to metamorphic rocks. In contrast, the easily weatherable quartz phyllite forms softer relief as can be seen in the right half of the photo at the 3,376 m high peak of the Sulden Spitze. Published with kind permission of © Dr. Alexander Stahr 2014. All Rights Reserved

Fig. 5.17 Mountain forms in the dolomites. In contrast to the Central Crystalline Alps, the Dolomites are characterized by often blocky massifs, massive rock towers, and vertical, several 100 m high walls. The name "Dolomites" for the South-Eastern Alps and "dolomite" for the rock and the mineral of which they are composed is the namesake of the mineralogist and geologist Déodat Guy Sylvian Tancred Grated de Dolomieu. In 1789, he pointed out the prevalence of a "with dilute hydrochloric acid non-effervescent rock in South Tyrol." In 1791, Horace Benedict de Saussure, one of the first people on top of Mont Blanc, dubbed the rock "dolomite" in his honor. The name was later transferred to the entire mountain range. Published with kind permission of © Dr. Alexander Stahr 2014. All Rights Reserved

Fig. 5.18 Mountain forms in the crystalline. Zillertal Alps, Austria. In the crystalline Central Alps ("crystalline" refers to metamorphic and plutonic rocks) landforms vary from hard rocks, resistant to weathering (granite or gneiss), to those susceptible to weathering. In the case of hard rocks, particularly in the glacial regions, peaks with steep faces and sharp ridges are common. Published with kind permission of © Dr. Alexander Stahr 2014. All Rights Reserved

Fig. 5.19 Formation due to structural evidence and rock features. The climate shaping of the mountains is affected by both occurring rocks and geological structure. In many cases, a separation of rock and structurally caused forming is hardly possible. The displayed peaks of the Drei Zinnen in the Sesto Dolomites consist of well-banked Dachstein Limestone or Haupt Dolomite. The Haupt Dolomite gives the rock a horizontal structure, while vertical fractures are responsible for the freestanding mountain forms. The base of the famous trio consists of more easily weatherable Raibl Beds that form the surface of a widespread plateau. The plateau itself is developed from Schlern Dolomite. The structure of the entire Drei Zinnen massif in its present appearance is thus based on the interaction of the different rock features, the geological structure, and processes of the climate-controlled weathering and erosion. Published with kind permission of © Wikipedia (public domain) 2014. All Rights Reserved

6 Shaping of High Mountains by Snow

A. Stahr and E. Langenscheidt, *Landforms of High Mountains*,
DOI 10.1007/978-3-642-53715-8_6, © Springer-Verlag Berlin Heidelberg 2015

Fig. 6.1 Snow patches. Sognefjell, Norway. In sheltered areas of periglacial altitude sometimes massive snow accumulations can be seen. They stay well until into the summer. Meltwater leads to water logging of the ground in daytime. At night there is the frost splitting effect. By the interplay of weathering and removal of residual detritus over time nivation hollows arise. Published with kind permission of

Snow is a major factor of actual landscape shaping in high mountains of the middle and high latitudes. A snow blanket demands the Earth's surface by pressure and movement, and meltwater also plays an important role. The process of landscape shaping by snow is called nivation (from Latin nix = snow). It can take place very slowly by frost splitting or gliding snow blankets, but also very quickly by snow slab avalanches.

6.1 Nivation Hollows

In sheltered areas mighty snow accumulations can last well into the summer. Through these snow patches flat molds of various sizes are developed. Some call these forms as nivation hollows. The flat floor of the nivation hollow is called nivation terrace.

In the transition area between a snow patch and rock the meltwater freezes at night. Therefore, frost splitting effects are here very intense. Residual detritus is carried away by meltwater and solifluction. This gradually creates a hollow form which increases to a nivation hollow. Even the origin of cirques begins with a snow accumulation in a terrain depression (see Chap. 8). Larger nivation hollows are also known as cryoplanation terraces (of Greek and Latin kryos = frost and planus = flat) (Figs. 6.1 and 6.2).

Fig. 6.2 Nivation hollows. Ötztal Alps, Austria. Nivation hollows are the result of frost slitting and removal of residual detritus in the area of snow patches. Published with kind permission of

6.2 Erosion by Moving Snow Blankets

On a relatively smooth surface snow blankets slowly slide down the slopes, on grass with long stalks. The rate of snow blankets' velocity depends on the nature of the snow and the steepness of the slope. It behaves like a viscous liquid. In addition to this the snow blanket creeps at speeds of a few

Fig. 6.3 Points for snow erosion. Berchtesgaden Alps, Germany. Natural obstacles such as boulders or young trees on steep meadows and pastures are ideal starting points for erosion by snow cover or ground avalanches. If they are not eliminated in meadows or pastures the moving snow takes them. This damage occurs at the soil surface. In order to prevent an expansion due to damage caused by further snow or water erosion, repairs to the affected areas are required. Published with kind permission of

millimeters to centimeters per day. Snow layers near the surface creep faster than layers near the bottom. Therefore, the result is a great tension within the snow blanket. If the limit of the load during creeping is achieved, it may lead to a break in the snow. The result is a basic avalanche. Snow blanket movements lead to the erosion of soil and belong to the process of mass abrasion.

The eroding effect of snow blanket is reinforced by carrying along materials such as stones, branches, or young trees. Erosion caused by snow is preferably there where the surface of the soil shows a certain roughness. This gives the snow blanket good weak points and is common in the periglacial altitude above the tree line with stony soils. Very large boulders on the other hand can prevent snow blanket movement, especially when the slope is covered by many large blocks in the upper third.

Affected by erosion or mass abrasion by snow blankets are particularly steep slopes with inclinations between 30° and 40°. Through this process, the surface is "shaved off" in layers: first the vegetation and humus layers, then the humic top soil, and finally the deeper parts of the ground. As a result, they are usually elongated, often several 100 m vegetation free areas. In English literature, these forms of erosion by snow blankets are often referred to as the term "Blaiken," a widely used term that comes from the European Alps.

Avalanches often follow predetermined hollow forms or trenches. We call these forms avalanche trenches. They are gradually expanded by annually recurring avalanches. At the high walls of the Himalaya and the Karakorum snow and ice avalanches fall several 1,000 m downhill. As a result of their high kinetic energy, they claim the rock very strong. Right wall canyons are usually the result of their eroding effect.

Snow, just as an avalanche, can also indirectly shape the landscape. If mountain streams may be dammed by large avalanches, a breakthrough of water can occur. This leads to catastrophic conditions in the valley (Figs. 6.3, 6.4, 6.5, 6.6, 6.7, 6.8, and 6.9).

Fig. 6.4 Block overlay. Karwendel Mountains, Germany. Woody plant groups or large boulders can increase the roughness of the surface and thus counteract snow movement. Published with kind permission of © Dr. Alexander Stahr 2014. All Rights Reserved

Fig. 6.5 Debris transfer. Berchtesgaden Alps, Germany. Moving snow covers lead to more or less pronounced shift of soil material. In conjunction with summer precipitation and rinsing it sometimes accumulates on the trunks of isolated trees or small groups of trees. The photo shows debris accumulation on the tree trunks with a conspicuous flexion like a saber. The snow every winter turns the young trees under its load. This strain leads to a curved growth at the lower range of the trunk. If the tree is strong enough to resist the pressure of snow, it grows above the curved section straight to the light again. Published with kind permission of © Dr. Alexander Stahr 2014. All Rights Reserved

Fig. 6.6 Erosion by snow (1). Bavarian Alps, Germany. Slipping or sliding snow covers can erode the soil extensively. The erosive effect is enhanced by the friction between rocks or branches carried along the soil surface. From this form of erosion stony soils are particularly affected. The eroded areas are irregularly limited with a significant expansion downhill. They can reach sizes up to several hundred square meters. Their depth is often only a few inches, sometimes several decimeters. These forms of erosion can occur in all high mountain areas with a strong snow cover in winter, especially at slope inclinations between 30° and 40°. Published with kind permission of © Dr. Alexander Stahr 2014. All Rights Reserved

Fig. 6.7 Erosion by snow (2). Berchtesgaden Alps, Germany. In loamy soils under grassland with greater depth of development snowslides or avalanches carry off about 20–50 cm thick soil layers including vegetation cover. The erosion forms show striking sharply defined ablation fronts with horseshoe-shaped curves. Their appearance is usually bound to slopes between 30° and 40°. This form of erosion is triggered by the compression of upper soil layers through snowslides or avalanches in waterlogged soils. Published with kind permission of © Dr. Alexander Stahr 2014. All Rights Reserved

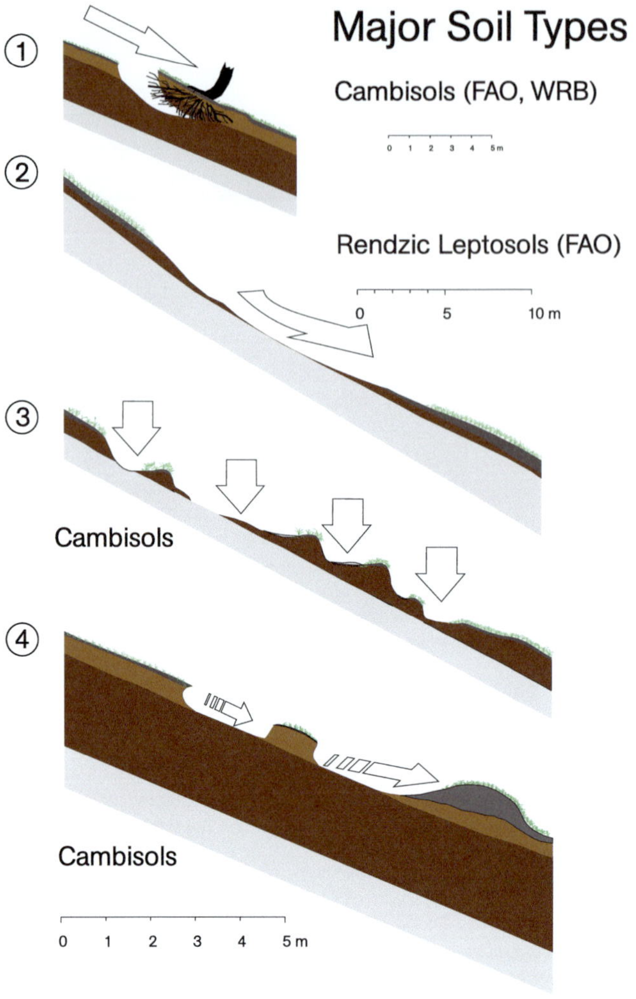

Fig. 6.8 Longitudinal profiles of typical forms of erosion by snow blankets and cattle path (schematically). 1 = snow erosion by pushing off young trees in cambisols, 2 = snow erosion in stony soils with little depth of development in rendzic leptosols, 3 = erosion by cattle path in cambisols, 4 = erosion form through snowdrifts and avalanches in deep developed soils (cambisols). Published with kind permission of © Dr. Alexander Stahr and © Dr. Ewald Langenscheidt 2014. All Rights Reserved

Fig. 6.9 Avalanche trench. Mühlsturzgraben, Berchtesgaden Alps, Germany. Avalanches often follow existing molds, for example, small valleys or trenches caused by tectonics. Published with kind permission of © Dr. Ewald Langenscheidt 2014. All Rights Reserved

7 Glaciers

A. Stahr and E. Langenscheidt, *Landforms of High Mountains*,
DOI 10.1007/978-3-642-53715-8_7, © Springer-Verlag Berlin Heidelberg 2015

When we think of high mountains, we have mostly glaciated peaks and mountain ranges in mind. And indeed there are many glaciers in high mountains of our globe as a significant landscape element. Many high mountains got their present appearance first in the recent geological time by flowing ice. In the Pleistocene, which began about 2.6 million years ago and ended about 11,600 years ago, ice age glacier streams have led to enormous changes. They gnawed on mountains, abrased the bottom of the valleys, and deposited large amounts of debris. Glaciers are a huge water reservoir for people who live in high mountains; their dammed meltwater supplies energy to millions, and without glaciers, high mountains would lose their charm and fascination for mountaineers, hikers, and nature lovers.

The name "glacier" was derived from the Latin word for ice "glacies." In French some write quite similarly: "glacier." In the Italian Alps it is called "ghiacciaio" and in the Rhaetian Alps "vedretta." The glaciers of Iceland are called "jökull" and the ones in Norway "bre." In Russian some says "lednik" and in Spain it is "glaciar" to them.

7.1 The Formation of Glaciers

Glaciers are made of ice and glacial ice is formed from snow. Apart from a few exceptions, there are anywhere on Earth glaciers above the climatic snow line, where the sun no longer melts away the solid precipitate. In Polar regions, this is already at sea level. Any fresh snow cover goes through a change after deposition. This process is called metamorphosis (from Greek metamorphóo = to transform). A distinction has been made between destructive and constructive metamorphosis. In destructive metamorphosis the branched structures of the hexagonal snow crystals are reduced by melting and evaporation. Branched crystals obtained therefore a grain shape and the snow gradually thickens. Fresh snow will gradually become old snow. If old snow outlasts an ablation period (from Latin ablatio = robbery) then it is called "firn" or "névé." The transformation of the snow is particularly accelerated by melting processes. It is therefore much more slower in polar regions than in warmer regions.

In case of constructive metamorphosis sublimation happens. It is the direct transition of snow crystals from solid to gaseous state. At this stage new crystals will be formed. They can be in the form of cup-shaped crystals, columns, or pyramid structures. Because of these results snow cover is called floating snow. Due to its loose storage in avalanche formation they play a crucial role. If the destructive metamorphosis progress goes further, individual components of the snow cover grow together. This leads to a stronger compression as well as to the decrease of pore volume and air permeability. Also in this sublimation plays a role besides thaw and freeze (= regelation). The firn gets impermeable as of a density of 0.80–0.85 g/cm^3. White firn ice forms are also called firnification. If firn ice gets into movement on an inclined underlay and at a further snow overlay, pressure variations arise. They cause further thaw and freeze. Through this the ice grains grow forward. Sublimation steps down since the air is still contained and squeezed out. Moreover the recrystallization in the ice works at a directional pressure. This is a moving of molecule layers in the ice crystal. Ice deforms and increases slowly also without melting processes. Finally firn ice develops to transparent glacier ice. This bluishly to greenish shimmering ice is air impermeable what included air bubbles still indicate. The density of the ice is 0.91 g/cm^3 now. For comparison: The density of fresh snow is about 0.1 g/cm^3. To form 1 cm of glacier ice, 80 cm of fresh snow is required. In addition to frozen water a glacier consists of different amounts of included mineral and organic materials like rubble, sand, dust and plant debris (Figs. 7.1, 7.2, and 7.3).

7.2 Properties of Glaciers

Glaciers are masses of ice that move under the influence of gravity. Ice flows at high altitudes, where the temperature is above freezing. There, ice is melting and the glacier is losing mass. A glacier is therefore divided into a zone of mass increase and a zone with mass loss. We call these zones nutrients and ablation zone (Figs. 7.4 and 7.5). The boundary between the two areas is the equilibrium line. On this line the accumulation area of a glacier is covered with firn, below of this it becomes exposed at summer so that the bare glacier ice comes to light. The equilibrium line is also called firn line, though sometimes there are differences between the two lines. Between firn line and equilibrium line a frozen melt water zone occurs. Some call frozen meltwater as superimposed ice. Large discrepancies between firn and equilibrium line are mainly found in subpolar glaciers.

7.2.1 Mass Balance

The difference between growth and mass loss results a net in the budget of the glacier. It determines whether a glacier retreats or moves forward. The extremely complex mass balance of a glacier can be simplified shown by the equation $b = c + a$. B is the mass balance, c the increased accumulation of snowfall, avalanches or frozen rain, and a is the ablation as a

Fig. 7.1 Old snow cover. Solar energy at high altitudes or in Polar regions is no longer sufficient to fully melt away the old snow in summer; with this the first step in formation of glacier ice has been fulfilled. Old snow that outlasts in sheltered areas above the climatic snow line or above the firn line of a glacier during an ablation period is called firn. Published with kind permission of © Dr. Ewald Langenscheidt 2014. All Rights Reserved

Fig. 7.2 Glacial ice. The individual components of a snow cover grow together by melting processes. This leads to greater compression with simultaneous decrease of pore volume and air permeability. This is known as sintering process of a snow cover. From a density of 0.80 to 0.85 g/cm^3 the snow cover turns into white firn ice. Through the process of recrystallization finally results transparent, greenish to bluish shimmering glacial ice. This ice has a density of 0.91 g/cm^3. As the process of metamorphosis extent on temperature, it takes place in Polar regions significantly slower than in warmer latitudes. For example, in Greenland the snow needs more than 100 years to turn into glacier ice. In the area of the Malaspina Seward System in Alaska, the conversion of snow to glacial ice only lasts 3–5 years. Published with kind permission of © Herbert Funk 2014. All Rights Reserved

Fig. 7.3 Impurities in glacier ice. Glaciers always contain different amounts of trapped mineral and organic materials such as rubble, sand, dust, and plant debris. Even dust from the Sahara is occasionally found in an alpine glacier. The photo shows that foreign material covers not only at the surface of the glacier as a thin superficial moraine, but also in the ice itself as internal moraine. Frequently mineral, but also organic material, reaches through crevasses into the glacier body. Published with kind permission of

result of melting, evaporation, wind drift or by calving glacier if a glacier flows into the sea, a lake or a fjord. Depending on whether accumulation or ablation dominates, it reduces or increases the mass of the glacier. If accumulation is greater than ablation, the mass balance is positive or otherwise negative. The time interval between two successive minima in the mass balance is the annual balance for the nets mass balance. Critical to nutrition of a glacier is the course of weather in the ablation period. Cool summers with snowfall in higher regions lead to mass increase. The winter temperatures and snow amounts are less important to the glacier behavior. However, even relatively mild winter temperatures with large amounts of snow favor to mass increase when a cool summer follows by heavy snowfall in the highest regions. In the tropics seasons are missing. Mostly on rainfall during rainy season is to determine the mass balance of glaciers. Ablation takes place every day. Because of the double rainy season in tropical glaciers even two annual balances may occur.

Glaciers in the middle latitudes and the tropics are warm or temperate with ice temperatures around the melting point. That they are not colder has the following reasons: the air temperature is often above freezing and by refreezing of melted water heat is released which contributes to increased temperature in the glacier. One gram of freezing water emits 80 cal to the surrounding ice.

Nevertheless warm glaciers become towards the depth more cold because with increased ice thickness the melting point decreases. Ice melts under pressure to a certain percentage. This change of state withdraws warmth from the ice. The temperature of ice then decreases in proportion to pressure. As in temperate glaciers there is always a proportion of water, they run all year meltwater which passes out in a glacier stream. Whether a glacier is "cold" or "warm" also depends on the altitude. Therefore, very high located alpine glaciers for example have regelation temperatures considerably under the melting point.

7.2.2 Why Does a Glacier Flow?

Glacial ice flows because of two reasons. First is internal deformation, known as plastic flow. And the second is sliding on a thin layer of water at the bottom of the glacier. High pressure inside the glacier leads to small shifts in the crystal lattice of ice which adds up to a movement of the entire ice mass. In addition to the plastic flow a glacier glides through regelation above its sole. At the bottom of the glacier the ice is under high pressure whereby the melting point of ice decreases. This is the same effect as it is with skating. The body weight on the narrow skate blade let the ice melt by pressure. The meltwater then acts as a lubricant on which the skater glides. Thus a glacier can also overflow obstacles in his bed by ice melting due from pressure of the obstacle and behind it freezes again. For glaciers in the Polar regions gliding through regelation is limited. Temperatures are so low that the glacier ice is frozen at the bedrock. Such cold glaciers move predominantly by plastic deformation or plastic flow.

The flow rate of a glacier increases from the edge to the center and from underground to surface. Also in vertical section a glacier shows a characteristic distribution of velocity. Apart from disturbances in icefalls, speed increases from accumulation zone until about the middle part of the glacier and then towards the end of the glacier decreases again. However, the velocity of glaciers which leads into the sea rises considerably at its tongue end.

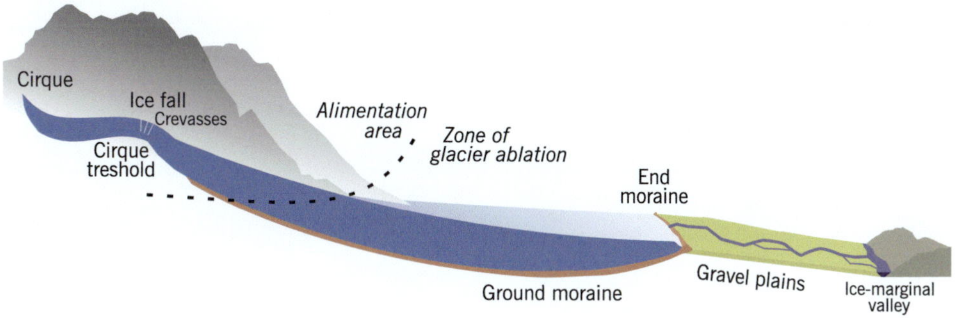

Fig. 7.4 Schematic vertical section of a glacier. Published with kind permission of © Dr. Ewald Langenscheidt 2014. All Rights Reserved

Fig. 7.5 Alimentation and depletion area. Pasterze, Hohe Tauern, Austria. Glaciers are divided into an alimentation (accumulation) and a depletion (ablation) area. The border between the two areas is the equilibrium line, which corresponds to the local or orographic snow line (*photo center*). Above equilibrium line a glacier gains in mass and below it melts slowly away again. Where the glacier in summer is covered with firn, there is the alimentation area. Below often follows sharply limited bare glacier ice, the ablation zone. This line is also called the firn line. In the same area, the firn line is significantly lower than the orographic snow line in the surrounding rocks. This is due to the cooling effect of the glacial ice. In the eastern Swiss Alps for example the difference amounts about 300 m and in the Caucasus even 800 m. Published with kind permission of © Dr. Alexander Stahr 2014. All Rights Reserved

Also the downhill gradient ultimately determines the flow rate. However, glaciers can be different from water flow and also can flow uphill if the firn thrust from accumulation zone is sufficient to push up a glacier. Alpine glaciers are moving normally at velocities of 30–150 m down into the valley annually. In the Karakorum and in Svalbard values were measured for the flow rate of glaciers by 130–800 m per year. Some glaciers on Greenland reach velocities of up to several kilometers per year.

7.2.3 Glacial Surge

A typical alpine glacier moves downhill about 50 m per year. After a longer period with lower velocity, it comes at some glaciers to a sudden and rapid movement. There are surge glaciers in all frozen areas of the world; however, most of these glaciers are in Alaska. Some have more than 200 ice streams; these are classified as surge glaciers. A surge of the 130 km long Hubbardglacier in the Gulf of Alaska in 1986

Fig. 7.6 Glacier tongue. Aletsch Glacier, Switzerland. A valley glacier can be divided into an alimentation and a depletion area. The latter includes the tongue, which can according to season and climatic conditions be collapsed to its end into itself or pushed up high. Glaciers, whose tongues are very steep or above an escarpment, carry a risk of ice falls. Published with kind permission of © Prof. Dr. Heinz Veit 2014. All Rights Reserved

caused tremendous changes to the ecosystem and led to enormous economic damage. The mechanisms of glacier surges are not yet entirely clear. However, a rapid glacier seems to precede an increase in water pressure in the melted water tunnels in the glacier ice. The glacier floats on the top and the friction at the bottom disappears almost completely. While the velocity of a glacier is usually the biggest in the middle and towards the edge steadily declining, it is barely changed in the surge condition from the middle of the ice up to its edge. A glacier surge ends by the outbreak of the accumulated water (Figs. 7.6, 7.7, 7.8, 7.9, and 7.10).

7.3 Ablation Forms on Glaciers

One of the striking ablation forms on glaciers are the impressive glacier tables. A large rock plate protects the underlying glacier ice from sun. Gradually an ice pillar arises under the top that is protected from the sun. Another very striking ablation forms are the penitentes ("nieve de los penitentes"). On steep icy walls in tropical high mountains you can find the so-called Riffelfirn or groove firn.

On stagnating glaciers, which does not move forward or still show a marked decline in the tongue, also appear abnormal ablation forms which are called thermal karst. The tongues of these glaciers are heavily covered with debris and surface moraine and have a number of hollow patterns and structures similar to lapies. Thus, they are in appearance reminiscent of a karstic limestone topography. This selective ablation is caused by different composition of overlying debris and by the internal structures of the ice. Parts of a glacier tongue without connecting to the main ice stream are called dead ice (Figs. 7.11, 7.12, 7.13, 7.14, and 7.15).

7.4 The Structure of Glaciers

Crevasses and stratifications are the predominant and best known morphological elements of the glacier's structure. In addition to this elements are shear planes. On the surface of a glacier, the pressure of ice is low in comparison with its base. The ice behaves depending on whether it is a tempered or polar glacier up to a depth of approximately 30–80 m as a brittle body. At places with great velocity differences of adjacent ice masses, it leads to crevassing if the upper area of the glacier will be dragged along by ice flowing plastically below. This happens primarily when the glacier moves about rounded hilltops or around curves. In the extreme case when a wildly torn ice fall arises the glacier is dissolved into real ice towers, the so-called séracs.

Cold Facts.
Glaciers stores about 75 % of the world's freshwater.

During the last ice age, glaciers covered 32 % of the total land area. Presently, 10 % of the land area is covered with glaciers.

(continued)

Fig. 7.7 Glacier calving. Roedefjord, Northeast Greenland. If glaciers end in the sea, in a fjord, or in a lake, the ice loss is mostly by calving, resulting in the known icebergs. The end of calving glaciers is known as calving front. The essential causes of the glacier calving can be seen in the buoyancy of ice sheets, the breaking of ice along zones of weakness such as transverse crevasses, the influence of gravity, or the force of the surf. The photo shows one of the many glaciers in Svalbard with its calving front into the sea. Published with kind permission of

Fig. 7.8 Glacier flows. Mont Blanc, France. In many high mountains glacier flows have a high economic importance. They are used in greenhouses for plant irrigation and they provide energy for powerhouses. At glaciers of the middle latitude as well as glaciers of the Arctic edge the glacier flows also in winter. This is partly due to the fact that by the regelation in temperate glaciers always a proportion of water is present. Furthermore, water that was stored during the ablation period is slowly released from cavities of the glacier. Published with kind permission of

Cold Facts (continued)

Antarctic ice is over 4,200 m thick in some areas.

Glacier ice crystals can grow to be as large as baseballs.

Due to the overlaying pressure glacial ice reacts with increasing depth much more plastic than near the surface. Therefore, a maximum depth of crevasses exists. In temperate glaciers crevasses reach depths of about 30 m. The intensity of the plastic response of the glacier ice is also dependent on temperature. Therefore, we find in the cold glaciers of Greenland or Antarctica crevasses of 80 m deep and more.

Crevasses are fixed and depend on the properties of the rock surface. Transverse crevasses occur where in the longitudinal profile of the glacier large velocity differences arise within the ice. After overcoming such a zone the crevasses close again. If the ice flows on steeper terrain transverse crevasses tear up wide and narrow to depth. We therefore speak of so-called V-crevasses, because they resemble the

Fig. 7.9 Glacier outlet. Glacier outlets are typical for tempered or warm glaciers. They occur also in polar glaciers during the snowmelt runoff in summer. Published with kind permission of © Herbert Funk 2014. All Rights Reserved

Fig. 7.10 Surge glaciers. Black Rapids Glacier, Alaska, USA. Some glaciers after a longer period with a lower velocity come suddenly to a fast movement. One describes such fast, often disastrous movements of valley glaciers as glacial surge. Most surge glaciers are in Alaska. The Black Rapids Glacier for example ventured in Alaska by a surge in the1930s within 5 months by 5 km; the 130 km long Hubbard Glacier provided disastrous conditions with a venture of 100 m/day at the Gulf of Alaska in 1986. Published with kind permission of © U.S. Geological Survey (USGS, public domain) 2014. All Rights Reserved

letter V in cross section. In hollows of the glacier ice is pressed. If it comes to rupture of transverse crevasses, they are in depth at the form of an inverted V or the letter A. This is referred to A-crevasses. Such a gap cross section, which is often described in textbooks as alpine reverse V-form crevasses, appears only in combination with low ice thickness. At big ice thickness such crevasses are very unlikely. At the end of a glacier, where the glacier tongue extends in all directions, fanlike radial crevasses result which look like a huge paw of a tiger or lion.

Longitudinal crevasses occur where differences exist across the glacier movement. This is for instance when the

Fig. 7.11 Glacier table. Ötztaler Alpen, Austria. The rock block protects the underlying ice from melting. The protected ice column therefore towers above the unprotected environment after long time. Published with kind permission of © Wikipedia (public domain) 2014. All Rights Reserved

ice flows over a convex ridge and parallel to the main direction of movement on its surface tearing columns arise. If the glacier flows over a cliff, it tears up in all directions. This causes cross crevasses that penetrate each other and they get a geometry which is dependent on the bedrock. As crevasses open up always perpendicular to the predominant tensile stresses, marginal crevasses are formed in the transition region from low to extremely high velocity. They are directed about 30°–45° against the glacier movement downwards and near to the edge of the glacier wide open and close with distance from the edge. The highest fixed crevasse is called "bergschrund." This crevasse occurs where the moving glacier ice separates from the stagnant ice above. "Bergschrund" is distinct from the "randkluft," which is not a crevasse but a gap on the border between snow and rock, the so-called black-and-white border. The randkluft is caused by the higher heating of the rock and the resulting heat to the ice.

Layering in glacier ice is formed in the accumulation area. It is a consequence of primary differences in the density of the snow cover. Winter stratums are packed looser than the water solid precipitations of summer. By melting processes, wind transports, or snow avalanches, fresh snow covers will be in addition redistributed. If all these layers will be cut on the surface, they can due to their different nature (winter or summer stratums, impurities, etc.) be seen as so-called ogives. In bare glacier ice millimeter to decimeter strong layers of white, air-rich ice changes to a bluish to greenish ice. This is known as white and blue bands. The location of these sheets shows a clear relationship to the pressure distribution in the glacier (Figs. 7.16, 7.17, 7.18, 7.19, 7.20, 7.21, 7.22, and 7.23).

7.5 Types of Glaciers

For typing of glaciers, certain properties or a combination of these are used. The following typing is based on the glaciers' situation in relief and their nutrition. This is one of the most established methods.

Glaciers may completely cover the relief or subordinate to this. The former belonging is the inland ice, ice sheets, plateau glaciers, and some ice caps of tropical volcanoes. The inland ice covered large areas in Greenland, Patagonia, and the Antarctic. It is the largest continuous ice cover of the Earth of today. The Vatnajökull in Iceland, an example of much smaller ice sheets, is the largest glacier in Europe and the largest glacier in the non-Arctic regions. One of the plateau glaciers is the Jostedalsbre in Norway; it is the largest glacier on the European continent. From all these types of glaciers numerous glacier tongues flow over the plateau edge into the deep.

In most of the high mountains valley glaciers are the predominant type of glaciers. They are subordinate to the relief. The longest of them is Lednik Fedcenko in the Pamir with a length of about 70 km. In Europe the Aletsch Glacier in Switzerland is the longest valley glacier with around 23 km in length. On the North American continent it is the Hubbard Glacier with a length of 122 km.

According to the type of accumulation (alimentation) area, the relief subordinated glaciers can be furthermore distinguished in firn stream glaciers, piedmont glaciers, névé basin glaciers, cirque glaciers, wall icing, hanging glaciers (ice aprons), glaciers formed by avalanches, and reconstituted glaciers. For firnstream glaciers the firn line is so deep that it still belongs to the glacier tongue. Some great mountain glaciers such as the Malaspina Glacier in Alaska in the foothills spread out in a fan. They are called piedmont glaciers. Glaciers, which flow over transfluence passes through whole valley systems, are known as ice stream systems. They were typical for the ice ages and are found today in Alaska and the island group of Svalbard (Figs. 7.24, 7.25, 7.26, 7.27, 7.28, 7.29, 7.30, 7.31, 7.32, 7.33, and 7.34).

7.6 Glacier Fluctuations

Glaciers have to adapt to environmental conditions. Cool climate phases cause an increase in mass of the glaciers. On the other hand, warm climate phases cause a loss of mass. If they persist long enough periods, the glaciers respond with a move forward or a melting back. We speak of glacier

Fig. 7.12 Thermal karst. Langthan Tsang Glacier, Nepal. Glacier tongues show sometimes striking ablation forms. Their surface is characterized by hollows and cauldron-shaped and karstic looking structures (*right in the photo*). This is called glacial or thermal karst (from Greek thermos = warm). This is due to a massive cover with debris and moraine. Through different levels of coverage with debris and moraine and different composition of the depositions, ablation acts selectively. Published with kind permission of

Fig. 7.13 Penitent snow. Andes, Chile. "Penitentes" or "Nieve de los Penitentes" are a striking manifestation of the ablation which primarily appears in tropical and subtropical high mountains. These bizarre forms remind of the shape of "penitentes" in their white shirts which travel around in processions in Spain nowadays during the Easter week. At the emergence of penitent snow the evaporation has a high quota. Furthermore, it is important that the ice or snow jags remain hard and dry during their development. They perfectly remain in subtropical high mountains with high solar energy, cold spells of fine weather, and low atmospheric humidity. Published with kind permission of

Fig. 7.14 "Riffelfirn." Cordillera Blanca, Peru. The "riffelfirn" or groove firn can be noticed in high mountain peaks of the subtropics and tropics mostly on steep icy walls. The tropic sun almost burns vertically down fascinating groove forms in the interaction with ascending damp airmasses from the jungles. The photo shows the southwest wall of the 5,947 m high Nevado Alpamayo. The Alpamayo is regarded as the most beautiful mountain of the world. Published with kind permission of © Wikipedia (public domain) 2014. All Rights Reserved

Fig. 7.15 Dead ice. The retreat of a glacier often affects parts of the glacier tongue that lose contact with the ice stream. They remain as immovable ice, as dead ice. This happened even with the ice age glaciers. For example, the famous Lake Chiemsee in front of the Bavarian Alps (the "Bavarian sea") has been created by dead ice. The Lake Chiemsee (*photo center*) has been saved from being buried by sediments until today; this indicates that for a long time a large mass of dead ice in the lake basin must have been present. Published with kind permission of © Dr. Alexander Stahr 2014. All Rights Reserved

Fig. 7.16 Glacier structures. The glacier structure is determined by structural or morphological elements like crevasses, stratification, or white and blue bands. Primarily stretch gaping crevasses and enormous ice towers in ice falls emerge as predominant forms. Published with kind permission of © Dr. Alexander Stahr 2014. All Rights Reserved

fluctuations. Seasonal, short-, medium-, and long-term fluctuations are distinguished. Seasonal glacier fluctuations are due to the change from winter accumulation and summer ablation period. The oscillations (from the Latin oscillatio = vibration) are therefore due to weather conditions. In the highlands of the inner tropics seasons are missing. Diurnal changes of weather will be largely balanced by the inertia of glacier ice. Oscillations are therefore hardly noticeable.

Short-, medium-, and long-term glacier fluctuations are caused by climate changes. An example of medium-term glacier fluctuations is the so-called little ice age, a cooling period from the beginning of the fifteenth century nearly to 1850. The alpine glaciers in the Middle ages were smaller than they are today, when in the seventeenth century a common cold relapse and a strong glacier advance occurred. The climax was reached in 1850. During this time, the average yearly temperatures were about 1 °C lower than today. Long-term glacier fluctuations by global climate changes led in the Pleistocene to several changes between ice ages (glacials) and warm stages (interglacials).

Since 1850 glaciers melt back with short interruptions. This trend is observed worldwide. U.S. scientists for example prognosticate a complete disappearance of the glaciers in Glacier National Park in the U.S. state of Montana within the next 70 years. Glaciers also change their appearance by glacier fluctuations. Advancing glaciers show a slight bulge in the accumulation area. These propagate later down on. On the other hand the end of the glacier tongues of melting back glaciers is flat and often their tongues appear disrupted.

The following photos show the glacier fluctuations in the glacier national park (Montana, USA) in a period of about 60 years at the example of the Grinnell Glacier (Figs. 7.35, 7.36, 7.37, 7.38, 7.39, and 7.40).

Fig. 7.17 Icefall. Mount Cook (Aoraki, 3,754 m), New Zealand. Icefalls are indicated by an extremely torn glacier body which flows down to the valley over a steep slope or an escarpment. The glacier is broken into needles and towers to an almost incalculable labyrinth from séracs and crevasses. Some find icefalls in all glaciated high mountains, provided that the relief shows the necessary minimum steepness of 25°–30° for the formation of an icefall. Published with kind permission of © Dr. Ewald Langenscheidt 2014. All Rights Reserved

Fig. 7.18 Séracs. In icefalls the glacier is frequently dissolved into single ice towers. Some describe these ice towers as séracs. They can get more than 20 m height and hold a high danger by their instability. Their collapse does seldom trigger ice avalanches. The name "sérac" became established in the scientific literature 1779 by the Swiss natural scientist and founder of the research alpinism Horace Bénédict de Saussure (1740–1799). He adopted the word "sérac" from the name of a soft cheese from the area of Chamonix (France). Published with kind permission of © Wikipedia (public domain) 2014. All Rights Reserved

Fig. 7.19 Crevasses. The velocity of a glacier is very different over this total area. It is determined by the shape of the rock surface. Therefore, it also has crevasses on a regular distribution. At steeper surface transverse crevasses occur. Longitudinal crevasses occur after narrowing, if it committed by a lateral pressure increased lateral movement. Published with kind permission of © Dr. Alexander Stahr 2014. All Rights Reserved

Fig. 7.20 Cross crevasses. If a glacier flows about a rounded hilltop, crevasses will arise which penetrate each other. These crevasses are called cross crevasses. Cross crevasses systems constitute the preliminary stage for an icefall. Published with kind permission of © Dr. Alexander Stahr 2014. All Rights Reserved

Fig. 7.21 "Bergschrund." Rötspitze (3,495 m), Venediger Gruppe, Hohe Tauern, Austria. The bergschrund is the highest stationary crevasse. This crevasse occurs where the moving glacier ice separates from the stagnant ice above. Published with kind permission of © Dr. Alexander Stahr 2014. All Rights Reserved

Fig. 7.22 Snowbridge. Large crevasses can be overcome by mountaineers only on snow bridges. A snow bridge should have been crossed in the early morning, because in the course of the day the snow becomes soft. Then there is the danger of breaking. Published with kind permission of © Herbert Funk 2014. All Rights Reserved

Fig. 7.23 Stratification. Stratification is an element of the glacial structure. It occurs in the accumulation zone of a glacier through the accumulation of solid precipitation. It results in a conspicuous sequence of different layers. Differences in the density of the snow cover, melting processes, input of organic materials, and mineral dust are the causes. Published with kind permission of

Fig. 7.24 Ice sheet. Ice sheets have a central nursing area from where the individual glacier tongues flow to different sides into valleys. The photo shows the Vatnajökull on Iceland. It is the largest glacier of Europe with an area of 8,456 km^2 and an ice thickness of up to 1,000 m. Published with kind permission of

Fig. 7.25 Plateau glacier. Jostedalsbre, Norway. It is the largest glacier in Europe with an area of 1,000 km^2. Published with kind permission of © Herbert Funk 2014. All Rights Reserved

Fig. 7.26 Icecap. Cerro Tapado (5,024 m), Andes, Chile. Icecaps are characteristic for many high volcanoes in the tropics. Published with kind permission of © Prof. Dr. Heinz Veit 2014. All Rights Reserved

Fig. 7.27 Valley glacier. New Zealand Alps. Valley glaciers are controlled by the relief. They are for most people the classical type of a glacier. Published with kind permission of © Dr. Ewald Langenscheidt 2014. All Rights Reserved

Fig. 7.28 Cirque glacier. Magdalena Fjord, Svalbard. In principle, this type of glacier is a relatively small valley glacier. In many cases a glacier tongue does not exist or is not developed significantly. Published with kind permission of © Herbert Funk 2014. All Rights Reserved

Fig. 7.29 Wall icing. Hohe Tauern, Austria. A glaciation of steep mountain flanks and walls is called flank or wall icing. On the photo you can see this form of glaciation at the 3,331 m high Fuscherkarkopf (*on the right side*). Characteristic for this glaciation is the fact that no typical névé basin exists. This happens frequently in the Himalayas, in the Karakorum, and in the Andes. Published with kind permission of © Dr. Alexander Stahr 2014. All Rights Reserved

Fig. 7.30 Hanging glacier. Rocky Mountains, Canada. A particular form of wall icing are hanging glaciers. They harbor the risk of catastrophic ice falls and ice avalanches. Published with kind permission of © Heiko Kurz 2014. All Rights Reserved

Fig. 7.31 Névé basin glaciers. At the névé basin glaciers the transformation of firn to ice takes place in a plain shape of the relief over the snow line. This can be a hollow, a depression, or a cirque (kar). Névé basin glaciers are the classic representative of valley glaciers. The photo shows the highest summit of the western High Tauern (Austria), the 3,674 m high Großvenediger (*picture center*) with typical névé basin glaciers. Published with kind permission of © Dr. Alexander Stahr 2014. All Rights Reserved

Fig. 7.32 Firn stream glaciers. Bering Glacier, Alaska, USA. Many of the glacier giants in Central Asia and Alaska belong to the type of the firn stream glacier. Characteristically for this type of valley glacier is the circumstance that a large portion of the nursing area still belongs to the glacier tongue. Published with kind permission of © U.S. Geological Survey (USGS, public domain) 2014. All Rights Reserved

Fig. 7.33 Avalanche basin glaciers. Blaueis, Hochkalter (2,607 m), Berchtesgaden Alps, Germany. Avalanche basin glaciers lay below the climatic snowline. Their nutrition happens exclusively by ice and snow avalanches. A known example in the Alps is the Blaueis at Hochkalter, the northernmost glacier of the Alps. Theoretically so-called reconstituted glaciers also can be classed as type of the avalanche basin glacier. This glacier type arises if a part of a glacier breaks off over a rock step and a new glacier is formed below the step of the fallen material. Published with kind permission of © Wikipedia (public domain) 2014. All Rights Reserved

Fig. 7.34 Ice stream system. Hornsund, Svalbard. Ice stream systems arise when valley glaciers are over passes in contact. The individual glacial streams feed on each other over the passes. In the quaternary ice ages such glacier systems were developed in many high mountains of the middle latitudes. Published with kind permission of © Herbert Funk 2014. All Rights Reserved

Fig. 7.35 Grinnell glacier 1938. Published with kind permission of © U.S. Geological Survey (USGS, public domain) 2014. All Rights Reserved

Fig. 7.36 Grinnell glacier 1981. Published with kind permission of © U.S. Geological Survey (USGS, public domain) 2014. All Rights Reserved

Fig. 7.37 Grinnell glacier 1998. Published with kind permission of

Fig. 7.38 Grinnell glacier 2009. Published with kind permission of

Fig. 7.39 Glacier moving forward. Jostedalsbre, Norway. Characteristic for a glacier moving forward is the more or less on-warped torn glacier tongue. The photo shows one of the numerous glacier tongues of the Jostedalsbre. Published with kind permission of © Herbert Funk 2014. All Rights Reserved

Fig. 7.40 Glacier going back. Denali National Park, Alaska, USA. The tongue of a glacier going back is comparatively flat. Parts of the glacier tongue are seldom at their retreat detached from the remaining ice body. These ice masses staying behind are called "dead ice." Published with kind permission of © U.S. Geological Survey (USGS, public domain) 2014. All Rights Reserved

Glaciers Form the Landscape

8

A. Stahr and E. Langenscheidt, *Landforms of High Mountains*,
DOI 10.1007/978-3-642-53715-8_8, © Springer-Verlag Berlin Heidelberg 2015

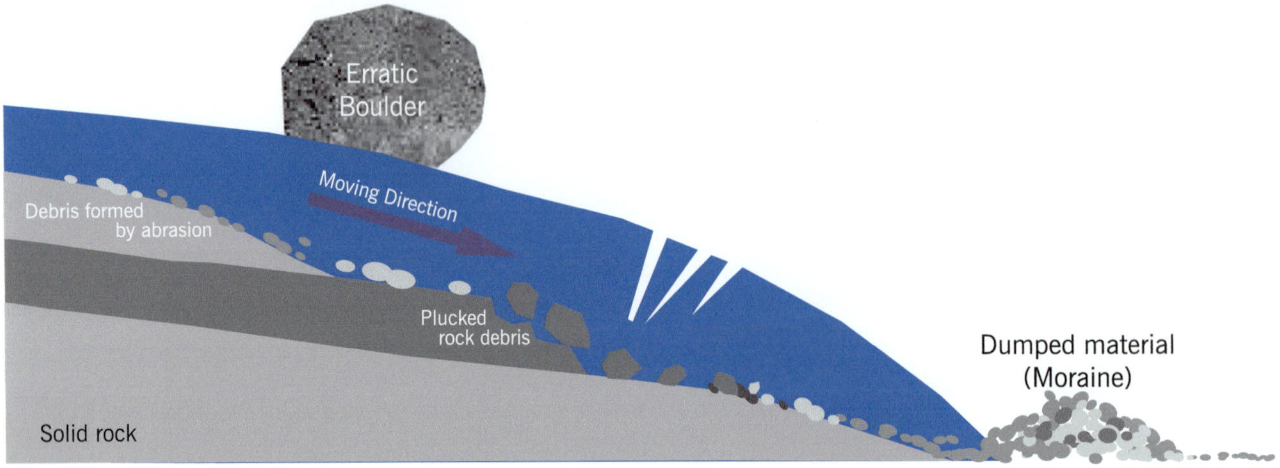

Fig. 8.1 Schematic figure of detersion, detraction, and exaration. Published with kind permission of © Dr. Ewald Langenscheidt 2014. All Rights Reserved

Glacial ice causes different processes of erosion and deposition. Rock debris of various sizes carried by the glacial ice stresses the bedrock. It grinds and polishes it. This process is called detersion (from the Latin deterere = to grind). Today the ice-free regions indicate by the polished and grinded rocks where and in what direction the glacier once flowed. Parts of the rocks at the base of a glacier which were frozen will break out by detraction (from the Latin detrahere = to tear off). Thereby bigger rock fragments can get out. This is called exaration (from the Latin exarare = to seam) (Fig. 8.1). On its front the glacier moves and compresses loose rocks. So a large number of characteristic landforms arise, which remain after the melting of the glaciers. Among these are cirques, U-shaped valleys, polished rocks, glacial striations, and different moraines.

8.1 Glacial Erosion Forms

If we study high mountainous regions formed by ice from above to below, we notice first the big armchair-like hollow molds in the rocks: the cirques (kar). Cirques are the shapes in the high mountain areas at which glaciers took and take their origin. The beginning of a cirque starts with a snow collection or snow spot in a depression on a slope. There, the snow lingers longer than in the surrounding area and the snow cover lasts well into the year. In colder climates, it is becoming a perennial snow patch (from Latin perennis = resistant). At its edges frost splitting processes are increased and the residual detritus is removed by the meltwater from this area. This gradually develops into a so-called nivation hollow or nivation cirque, the prestage for a cirque. With increasing snow thickness a snow patch becomes more and more similar to a glacier and over time glacier ice arises out of the snow patch. When the surrounding rock formations are eroded extensively, finally there is a typical cirque, a bowl-shaped depression carved out of a mountain.

The glacier ice is pressed and erodes to the middle of a cirque. Thus a typical cirque threshold is formed towards the valley side. After ice melting a cirque tarn can often be found behind the cirque threshold. The cirque threshold itself is polished by glacial ice and reveals distinct glacial scratches, unless it is covered by moraine. By continuing glacial erosion several cirques can grow together whereby high rocky ridges (aretes) can develop. During the ice ages innumerable mountains of the Alps, the Himalayas, and other high mountains were surrounded on all sides by glaciers. After the melting of the glaciers, mountains remain, which are today completely bounded by high rocky ridges. Such shaped mountains are known as pyramidal peaks or glacial horns (karling). During the ice age glaciers overflowed even high-altitude passes. So we find today in high mountain ridges U-shaped incisions, the so-called transfluence passes. They show the former crossing of a glacier into a neighboring valley. From the glacier ice completely overflowed rock formations are to be redesigned as so-called sheepbacks or mamillated rocks (in French "roches moutonnées" = mutton backs).

Below the cirque threshold valley glaciers follow the course of valleys, which had already been preglacial by tectonic processes and streams. As a result of detersion and detraction, formerly V-shaped valleys have been transformed into U-shaped valleys by deep and lateral erosion. The rocky slopes of these valleys are generally very steep and have a rounded cross section at its base.

In addition to the processes of detersion and detraction during glacial erosion, glacial meltwater also plays an important role. Because of the high velocity of under high hydrostatic pressure standing melted water in a glacier, the process of cavitation is effective (from the Latin cavare = to cave).

Fig. 8.2 Glacial high mountain landscape. Fitz Roy (3,406 m), Andes, Argentina. High mountain regions characterized by glaciers exert an irresistible stimulus on us by their form variety in a confined space. Their curt rocky ridges and deep valleys cut by the ice of the gnawing glaciers draw millions of people worldwide into their spell every year. Published with kind permission of © Wikipedia (public domain) 2014. All Rights Reserved

Fig. 8.3 Glacial polish. Glacier garden near Inzell, Bavaria, Germany. Landscape areas in the high mountains which were glaciated formerly show striking rock surfaces. Because the rocks seem to be really polished, some speak about the glacial polish. These striking surfaces got well preserved under today's snowline or in areas where the ice have disappeared only in few years or some decades and where other erosion processes had not destroyed them until now. Published with kind permission of © Dr. Alexander Stahr 2014. All Rights Reserved

If water flows very fast, its static pressure decreases so far that interior steam-filled cavities were formed. When the flow rate drops down behind a barrier, the water bubbles implode with great force. This creates shock waves that erode the rock in the lee of the meltwater stream such as hits with a hammer. However, this process is linear and together with detersion and detraction a U-shaped trough valley can be created.

Side valleys, leading to a U-shaped main valley, are often situated high above the valley floor of the main valley. This phenomenon is due to the fact that the erosive power of smaller glaciers was in the side valleys lower than in the main valley.

In the longitudinal the glacially overprinted main valley may also have significant escarpments. Therefore, two reasons come into consideration. The first is that at the confluence of two glaciers the ice pressure on the ground was intensified

Fig. 8.4 Glacial strias. Glacier garden near Inzell, Bavaria, Germany. Blocks at the base of a glacier cannot be pressed straight up into the ice. They are rather dragged along by the glacier and scratch a long groove into the bedrock. The block itself is grounding down flatly. Therefore the most prominent features of glacial erosion and dispatch of rock material by a glacier are the strias and weals on the solid rock floor. Published with kind permission of © Dr. Alexander Stahr 2014. All Rights Reserved

Fig. 8.5 Cirque (Kar). Hochkalter (2,607 m), Berchtesgaden Alps, Germany. A cirque is the place of the beginning of a glacier. Superficially on a slope snow will be gradually collected and stays under favorable climatological conditions over the whole year. From there melted water leads the fine weathered rubble away which permanently arises by frost action. If ice will be formed gradually in the course of time from snow, erosion will be strengthened. From a primary nivation niche in the rock a hollow form arises, which is often described as a cirque. The matured cirque is surrounded by steep walls in the back and on the two sides. The cirque ground is rebored in a trough shape by the ice which pushed from all sides to the middle of the cirque. A cirque is often completed by a cirque threshold towards the valley. The cirque threshold occasionally can be covered with moraine after the melting of the ice. Published with kind permission of © Dr. Ewald Langenscheidt 2014. All Rights Reserved

Fig. 8.6 Cirque lake. Hohe Tauern, Austria. A cirque lake frequently arises behind a cirque threshold after the retreat of the ice. The lake is damed by the rock bolt of the cirque threshold or by moraine material remained on it. The picturesque cirque lakes are a center of attraction and walking destination of first rank for tourists, mountain hikers and relaxation seekers in many high mountains. Published with kind permission of © Dr. Alexander Stahr 2014. All Rights Reserved

Fig. 8.7 Rocky ridge. By the gradual merging of adjacent cirques, the ice is shaping sharp rocky ridges. At the shown Matterhorn (4,477 m, Switzerland) four enormous rocky ridges appear, which represent famous ascent routes: Furggengrat, Hörnligrat, Zmuttgrat, and Liongrat. Published with kind permission of © Wikipedia (public domain) 2014. All Rights Reserved

locally. Second, a valley glacier had to overcome steep passages along the path of the valley that existed prior to the glaciation. There the glacier moves faster and exerts higher pressure on the valley floor and the valley is there more eroded. The result of these processes is the characteristic stepped longitudinal profile of a glaciated valley.

In the pleistocene ice ages glaciers reached the flood plains of the mountains. Here too, the ice followed molds or existing zones of weakness. With the retreat of the ice, the molds were filled with meltwater. This happened at the so-called finger lakes, for example, at the 538.5 km^2 wide Lake Konstanz (Figs. 8.2, 8.3, 8.4, 8.5, 8.6, 8.7, 8.8, 8.9, 8.10, 8.11, 8.12, and 8.13).

8.2 Glacial Deposits

Rock fragments which are taken, transported, and matured by the glacier are described as a moraine. In principle terminal moraine, shore moraine, lateral moraine, middle moraine, inside moraine, sub-moraine, ground moraine, and top moraine are distinguished. There are more exotic

Fig. 8.8 Karling. Königsspitze (3,859 m), Ortlergruppe, Italy. Karlings are mountains which are limited by sharp rocky ridges. They were "gnawed off" by ice from all sides. Gradually rugged peaks arise from rather gently formed massifs by glacial erosion and cirque formation. Published with kind permission of

Fig. 8.9 Trough valley. Berchtesgaden Alps, Germany. Steep valley flanks and a rounded cross section at its base are typical for a trough valley. Therefore, some also describe a trough valley as a U-shaped valley. Trough valleys have resulted from former V-shaped carved valleys arisen from running water by later glacial erosion. Published with kind permission of

moraine types like dam moraines and pedestal moraines. The most of moraine material consists of different sizes of an unsorted mixture of sand, gravel, earth, loam, and other rock debris. The components are a result of detersion, detraction, and exaration or grazed by weathering from the glacier's surrounding rock faces and reached from there to the glacier by avalanche transport for example.

Terminal moraines also as forehead moraines are described from prominent embankments at the glacier forehead. These embankments as a full molding stand out

Fig. 8.10 Trough end. Pasterze, Glocknergruppe, Austria. The valley end of a glacial trough valley is frequently formed by a rock step. This trough end is the area of the valley in which the valley shaping in front of icing-up by flowing water starts more and more. Published with kind permission of © Dr. Alexander Stahr 2014. All Rights Reserved

Fig. 8.11 Transfluence pass. Chiemgau Alps, Germany. During the ice age ice masses flowed over lower ridges from one valley to another. Today, similar phenomenons can be seen in Spitsbergen, for example, the overflowed area or ridges lay higher than the present glaciation. Some describe these cuts between different valleys as transfluence passes (from Latin trans = over and fluere = to flow). Published with kind permission of © Dr. Ewald Langenscheidt 2014. All Rights Reserved

Fig. 8.12 Hanging valley. Venediger Gruppe, Hohe Tauern, Austria. Hanging valleys can be seen at numerous places in formerly glaciated high mountains. Smaller valleys lead to the main valley far above the valley ground of nowadays. The cause for these partly very high area steps is that the main valley was filled with a large glacier which deepened the main valley more than the smaller side glaciers the tributary valleys. There are now often gorges and waterfalls between the high situated side valleys and the main valley floor. Today, roads of the main valley often wind upwards to the high-altitude valleys over the steep step in double bends. Published with kind permission of © Dr. Alexander Stahr 2014. All Rights Reserved

Fig. 8.13 Fjord. Geirangerfjord, Norway. A fjord is a trough valley which was flooded by the sea. Fjords therefore look like sea bays intruding into the country deeply with extremely steep rock faces. A fjord is characteristic for its lowest water depth into the open sea shortly before the muzzle. This threshold results from the glacial deepening of the valley like a corrie threshold. Ice age fjords are typical landforms of the Norwegian west coast. The deepest fjord outside of the Antarctic is the 1,308 m deep Sognefjord in Norway. It extends 204 km far into the country with a depth of 847 m at its beginning. The deepest fjord of the earth is the 2,287 m deep Antarctic Vanderfjord. Distinctive fjords are also found in Scotland, called firth there, and in Alaska, Patagonia, and New Zealand. One of the well-known fjords of New Zealand is the Milford Sound in the New Zealand Alps. Published with kind permission of © Herbert Funk 2014. All Rights Reserved

Fig. 8.14 Terminal moraine. Waging am See, Chiemgau, Germany. Terminal moraines rise as arched embankments at the glacier forehead. Its emergence is carried out at pushing forward or stagnating glaciers. At standstill of the glacier forehead a stacking moraine will be formed and at ventures a compressing moraine. The photo shows a terminal or embankment moraine. The striking embankment is the area of the wooded mountain range. Published with kind permission of

strikingly from the other area surface after melting of the ice and show the expansion of the ice. The terminal moraine consists of a conglomeration of material from inside moraine, top moraine, ground moraine, and sub-moraine. The inside moraine is nothing else but rock material which got inside the ice by crevasses for example. A top moraine is found primarily at avalanche-fed glaciers or at glaciers flowing beside high rock faces. Avalanches always deliver fresh material which covers the glacier surface in different thickness. Therefore sharp-edged forms of the rubble are also typical for the top moraine. So a thick top moraine protects the ice from melting and covered glacier tongues reach deeper parts of a valley as the bare ones.

The ground moraine consists mostly of rock fragments of different sizes which were matured at the bottom of the glacier or grazed by glacial erosion. The larger components of the ground moraine usually show well-rounded edges in contrast to the top moraine. Sub-moraine describes frozen rock material at the base of a glacier. Because of the unclear separation to the ground moraine this concept is used, however, hardly or only seldom. When melting, the top moraine of the glacier tongue, but also the inside moraine, is gradually put down on the ground moraine. The difference between these moraine depositions in a sequence can be seen easily because the ground moraine contains more fine substances and is more compressed due to the ice pressure.

Terminal moraines can be classified in different types and to be more precise into stacking moraines and compressing moraines. The appearance of the two types depends on the mass household of a glacier. Where the glacier tongue stays or stagnates for some time, stacking moraines form themselves. Melting processes and the glacier advance are in a scale. On warm summer's days similar to a conveyor belt top moraine material dumps over the steep tongue end and gradually builds up the final moraine embankment. Additionally material from the inside moraine and partially debris of the ground moraine join the deposition at the base of the embankment. If the mass balance of the glacier is positive, it compresses the terminal moraine like a bulldozer. Differently old retreat stadia of a glacier are marked by a sequence of several final moraine embankments. Some speak about echelon-like terminal moraines.

The debris transported and moved at the edge of a glacier forms the side and shore moraines. These concepts are often used synonymously; however, it is not quite correct. The lateral moraine is a special shape of the top moraine and therefore basically the rubble material is transported at the glacier edge. Only if the rubble material of the lateral moraine accumulates at the ice edge or is removed at this, the prominent shore moraine forms itself. At the confluence of two glaciers the respectively facing lateral moraines become a middle moraine. Arising from the layer of the meeting moraine material within the glacier this middle moraine is therefore just a special top moraine. One can watch very beautifully the middle moraines at the large glacier streams of the Himalayas, the Karakorum, or Alaska. The Aletschgletscher in Switzerland shows for example typical sustained middle moraines. A shore moraine minted

Fig. 8.15 Lateral moraine. Mer de Glace, Mont Blanc area, France. In the picture a broad stripe can be discerned from lateral moraine. One can see also a narrow rubble tape, the middle moraine in the middle of the glacier. In the background the enormous north face of the 4,208 m high Grandes Jorasses rises over the Mer de Glace. Published with kind permission of © Dr. Alexander Stahr 2014. All Rights Reserved

Fig. 8.16 Shore moraine. Rocky Mountains, above lake Louise, Alberta, Canada. The glacier's rubble of the sides of the top moraine melted from or removed at the side forms the embankment of the shore moraine (*moving into picture center from the left*). The over-stamp in an older shore moraine by a renewed glacier venture also can lead to the stratification. Published with kind permission of © Wikipedia (public domain) 2014. All Rights Reserved

Fig. 8.17 Medial moraine. Wortmanns Glacier, Chugach Mountains, Alaska. A medial moraine (*photo center*) arises at the confluence of two adjacent glacier streams from the meeting lateral moraines. The top moraine is at the same time also the medial moraine; it depends on their position at the glacier's surface. Like every top moraine the medial moraine is also bound to the snow line. Published with kind permission of © U.S. Geological Survey (USGS, public domain) 2014. All Rights Reserved

well and seeming fresh appears at many glaciers of the Alps at the glacier front end as a striking land form. It indicates the height of the glaciers at the middle of the nineteenth century. Until now they have melted back with only short interruptions from there.

Dam moraines are typical for subtropical high mountains. These moraines in principle are nothing else but thick ground moraines on which a glacier drains away like on a railway embankment. Avalanches departure from high steep faces and it is necessary to have an intensive and frequent

Fig. 8.18 Inside moraine. Renbugt, North-East Greenland. Inside moraine is the rubble transported within the glacier. It gets into the ice by crevasses or by processes of glaze and freezing at the glacier's surface. The inside moraine therefore consists of material of the top moraine which gets due to slope erosion and avalanches onto the glacier. In the area of rock spurs which rise from the glacier inside moraines can also be subglacially formed by detraction. The rubble material pulled out at the rock spur by the glacier is surrounded and taken away by the ice. Published with kind permission of © Herbert Funk 2014. All Rights Reserved

subtropical frost action for the formation of this ground moraine. Therefore, it can be found particularly in high mountains like the Himalayas.

Beautiful examples of pedestal moraines can be seen in the Himalayas, in the Andes, or at Greenland. This moraine type is in theory also a special kind of ground moraine. Its emergence is explained by the main valley shore moraine or the main glacier itself on the over-string pulling of a hanging valley glacier. In the second case, an inter-moraine which gets still more thick with the sinking of the surface of the main valley glacier will be formed between two ice bodies first. After melting of the push-aside glacier a moraine pedestal remains which extends from the main valley ground up to the bottom of the hanging valley. Very impressive moraines of this type can be seen at the foot of the 4,500 m high Rupal flank of Nanga Parbat (Himalayas).

Occasionally very big single rock blocks can be found in landscape areas of high mountains glaciated formerly which can be strikingly different from the rock of the surrounding area. They actually go astray because they were partially transported and removed together with moraines over wide distances and deposited after the melting of the ice. Erratic boulders got therefore most of more or less smooth rounded corners. These rock blocks basically are the greatest grain sizes which were transported by the glaciers.

Drumlins represent a strange form made of ground moraine material (of Irish druim = a small back). It is a streamlined glacial deposit form of elliptical ground plan. A drumlin shows an exactly reversed form like a round hump in longitudinal section. Drumlins happen not alone but in whole groups. Drumlinfields are found primarily in the forelands of high mountains. A famous example from the foothills of the Alps, southwest to Munich, is the drumlinfield between Lake Ammer and Lake Starnberg. Some explain the subglacial forming of the Drumlins, as a forming under ice by reduced ice speed in the inner periphery of a glacier. Until now, there is no study about how a drumlin arises exactly (Figs. 8.14, 8.15, 8.16, 8.17, 8.18, 8.19, 8.20, 8.21, and 8.22).

Fig. 8.19 Ground moraine. Hohe Tauern, Austria. The components of the ground moraine have usually smoothly rounded edges; greater components have a clear lineation often recognized on their surface. This debris is stored thick because of the high compressive stress and consists of rougher components in a finer matrix of loam. Through crushing at the transport loam arises of the rubble. Some speak about debris marl at existing lime quotas. The photo shows a ground moraine with blocks in a finer matrix. Published with kind permission of © Dr. Alexander Stahr 2014. All Rights Reserved

Fig. 8.20 Surface moraine. Aletsch glacier, Switzerland. Rock rubble on the glacier's surface forms the top moraine. Surface moraine will be formed intensively under steep rock skirts. The material of the top moraine therefore consists of sharp-edged rubble unlike the material of the ground moraine. If the material of the top moraine reaches the ice mass by columns or glaze processes, it becomes an inside moraine. It builds up step by step the terminal moraine at the tongue end of a stagnating glacier. The tongue of a melting glacier sinks gradually into itself. After its complete disappearance the surface moraine stores on the ground moraine and it is then called ablation moraine. Published with kind permission of © Prof. Dr. Heinz Veit 2014. All Rights Reserved

Fig. 8.21 Erratic boulder. Rannoch Moor, Northwest Highlands, Scotland. Occasionally very big single rock blocks can be found in high mountains or forelands which are completely different from the rocks of the immediate surroundings. These are called erratic blocks or boulders (of Latin errare = go astray). Erratic blocks were partly transported from far distances and dropped after the melting of the glacier. They are consequently more or less rounded and show clearly scratches which are due to the moraine material moved along with them. These rock blocks often have diameters of several meters. Published with kind permission of © Dr. Alexander Stahr 2014. All Rights Reserved

Fig. 8.22 Drumlin. Upper Bavaria, Lampoldinger Drumlinfield, Germany. Drumlins are full forms with an elliptical ground plan which consist of ground moraine material. Steeply the sides turned towards the ice venture or turned away long and declined. Drumlins usually appear in groups. Extensive drumlinfields are found primarily in the forelands of high mountains. Published with kind permission of © Dr. Ewald Langenscheidt 2014. All Rights Reserved

9 Meltwater and Landscape

A. Stahr and E. Langenscheidt, *Landforms of High Mountains*,
DOI 10.1007/978-3-642-53715-8_9, © Springer-Verlag Berlin Heidelberg 2015

Fig. 9.1 Meltwater. Briksdals breen, Norway. Meltwater from the glacier is part in the shaping of high mountains. Published with kind permission of © Herbert Funk 2014. All Rights Reserved

In addition to the glacial processes meltwater forms the landscape as well. It also transports and deposits rock materials in different particle sizes. It erodes rocks in different ways and is flattening rock fragments, which themselves provide rock abrasion. The melted water works hand in hand with the glaciers. Some speak therefore of "fluvioglacial" erosion and deposition. If these processes are happening outside of the glacier instead, it is called "glacialfluvial." In scientific literature, there is often the synonymous use of the two terms (Fig. 9.1).

Conspicuous fluvioglacial erosion forms are the potholes. The strange holes, looking like wells in the solid rock, excite the imagination of people for centuries. There has been no scientific explanation for a long time. Thus, these phenomena were considered as the work of witches and giants.

It took a long time until these forms have been associated with the glaciation in context. In the nineteenth research century, the Swiss-American geologist, zoologist, and paleontologist Jean Louis Rodolphe Agassiz (1807–1873), the Swiss geologist Albert Heim (1849–1937), the German geographer and meteorologist Eduard Brückner (1862–1927), and the German geographer Albrecht Penck (1858–1945) had caused the fundamentals.

Potholes are the results of abrasive action of rock fragments. In fair weather in summer, the glacial meltwater collects in rivulets and rivers on the ice surface. Where rip columns exist, it disappears into the depth. In this so-called moulins, vertical tubes and ducts are formed in the ice by the tumbling down water. The water also leads with it sand and pebbles. In some spots at the underground of a glacier turbulently flowing water begins to deepen into the bedrock. Comparable to a sandblaster, the battered rock materials are thrown with speeds up to 200 km/h against the wall of the resulting glacier pot. The results are horizontal potholes in the bedrock or lateral forms at rock walls. Depending on the local conditions of origin, they can reach diameters from a few centimeters up to more than 10 m. Smaller molds of this kind can sometimes be produced within a few weeks or one summer. Perched above the valley floor potholes that were grounded into the side of cliffs testify to a powerful glaciofluvial ice cover in the valley. Similar forms are the potholes, which can be observed along valley walls of rivers (Fig. 9.2).

Fig. 9.2 Pothole. Reit im Winkl, Chiemgau Alps, Germany. Potholes are the result of abrasive action of rock fragments. The results are horizontal potholes in the bedrock or lateral forms at rock walls. Published with kind permission of © Dr. Ewald Langenscheidt 2014. All Rights Reserved

9.1 Fluvioglacial and Glacialfluvial Deposits

In the proximate foreland of high mountain glaciers, but also further down the valley in the glaciated terrain, there are a variety of landforms, resulting in whole or in part of melt-water deposits. It is not always easy to follow the complex processes of the respective forms or deposits in the area. For in the development of many types of terrain in the glacier environment affects only the transport and deposition by meltwater and gravitational processes in the form of rock or debris flows.

In addition, meltwater deposits are often interlinked with glacial deposited moraines. Numerous deposits and landforms in the glacier foreland are therefore polygenetic phenomena. In principle, it should be noted that fluvioglacial or glacialfluvial processes can happen on the surface of the glacier, in the front of a glacier, and at the edge of the glacier tongue.

The reason a crevasse infilling may have a sediment thickness of the upper and inner moraines is that it has been deposited in columns by rushing meltwater. However, this requires relatively flat terrain at the column base and a slowly melting glacier or even dead ice. In such a situation some speak also of stagnant ice. After the complete melting of the ice back the fluvioglacial sediment infilling remains as a isolated solid form on the ground moraine, which is called kame (from the Scottish word kame = steep hill of loose material).

As a result of deposition by flowing water such forms usually have a visible material sorting. A restriction with respect to a visible stratification makes the material turn over after the melting of the surrounding ice, and then it is stored chaotically. In case of a stagnant or even advancing glacier, the sediment can be cleared again, in steep terrain to not significant deposits at the column base. In this manner sedimented material is relatively rare in high mountains and often only visible in small areas. This is due to the steepness of the terrain and the quick removing by streams at the valley bottom. In contrast, such sediment accumulations at the column bases are well preserved in territories of former lowland inland ice sheets, such as in southern Scandinavia. They extend like a wound over the ground moraine, sometimes for miles and are referred to as "os."

Striking are terraced deposits along the sides of the valley, the kames terraces. At the border of the ice significant levels of sediments can be built up. Such steps or terraces, which have been preserved due to erosion and fragmentation often only as fragments, are sometimes reaching heights of several tens of meters. They are composed of wall rubble, gravel, and sand and they are often layered, which refers to the formation as glacialfluvial sediments.

An essential process to form kames terraces runs over the way of the development of sander banks. In this case, the meltwater flows between lateral moraine and a valley wall from meandering. The sedimented material shows a clear stratification and sorting. Melts the ice, the remaining terrace consists of lateral moraine at the valley side and towards the flank of glaciofluvial sediments. Often the lateral moraine is eroded by undercutting, so that only the sander material remains. There is also the possibility that the sander is poured directly between glacier and valley wall.

Also gravel, when the melted water poured it into columns, looks like kames terraces after thawing of the glacier. Therefore, it is often difficult to understand the actual process of creating a glaciofluvial form — kames terraces for example can also arise in dead ice fields. To the aforementioned formation conditions of kames terraces, there also exists the material supply from valley slopes, whether as rockfall, avalanche debris or mudflow. By gravitational processes the ice edge materials fed interlock or mixed with other glacialfluvial sediments. This points to the problem of reconstructing the exact formation process of a bank formation. It is a fact that often the deposited material after the end of the ice age is turned over or obstructed by slumping. It should be mentioned that also drops from superficial moraines or glaciofluvial sediments on the ice bodies can leave locally rubble mounts.

In front of the exit points of glacial rivers out of the glacier tongue, larger shallow sediment compartments are poured as gravel plains and flood plains; they are called "sander" in the alpine region. From the glacier front to the outer ranges of the gravel floor is a particle sizing from coarse gravel to gravel and to fine sands. Even finer particles will be transported further on by glacial streams. Within the gravel floor the glacial stream runs frequently in many branched channels or is jammed back into a lake (Sandersee). Sander occupies the largest area of fluvioglacial or glacialfluvial sediment deposits (Figs. 9.3, 9.4, 9.5, 9.6, and 9.7).

Fig. 9.3 Kame. Seeoner Lakeland, Chiemgau, Germany. The well-defined full form in the *left center of the photo* shows a kame north of Lake Chiemsee in the Seeoner dead ice field. Published with kind permission of © Dr. Ewald Langenscheidt 2014. All Rights Reserved

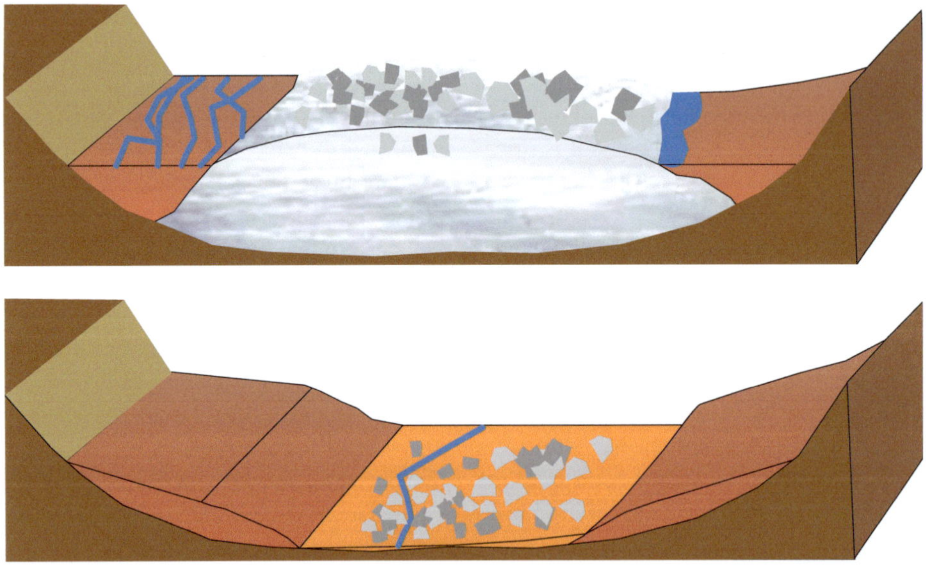

Fig. 9.4 The formation of kames terraces. Kames terraces are striking full forms, which originated out of glacialfluvial deposition of sand and gravel between glacier and valley wall. In general, the sediment of kames terraces is layered very thick. Kames terraces can reach heights of 20 m and more. Published with kind permission of © Dr. Ewald Langenscheidt 2014. All Rights Reserved

Fig. 9.5 Gravel plain. Spitsbergen, Norway. Gravel plains are gravel deposits that were sedimented by the glacial meltwater runoff in front of the ice edge. They are attached to the glacier, where they are discharged from the glacial stream poured into a fan-shaped apron. The sequence of ground moraine, terminal moraine, and gravel plain or sander was first described by the German geographer Albrecht Penck (1858–1945). Published with kind permission of © Herbert Funk 2014.

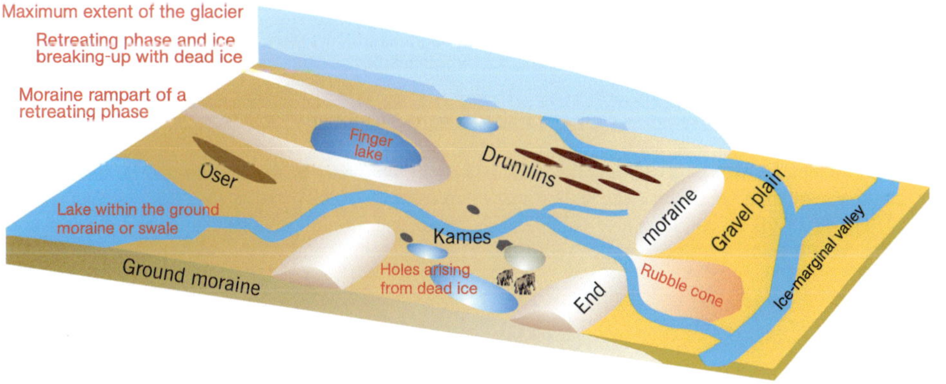

Fig. 9.6 The sequence of glacial deposits. Published with kind permission of © Dr. Ewald Langenscheidt 2014.

Fig. 9.7 Kames terraces. Wipptal, Austria. Between the lateral moraine of the ice body or a retreating glacier and the valley wall, meltwater gravel, gravel, and sand bank up and form a so-called bank sander. If the glacier melts, there remains a striking strip of sediments from the glacial edges and the lateral moraine, which is called kame terrace. Typically, these glacialfluvial incurred relief forms show in their interior a clear stratification and material sorting. Published with kind permission of © Dr. Alexander Stahr 2014. All Rights Reserved

10 Water and High Mountains

A. Stahr and E. Langenscheidt, *Landforms of High Mountains*,
DOI 10.1007/978-3-642-53715-8_10, © Springer-Verlag Berlin Heidelberg 2015

As a landscape element, water is shaping the mountains at all altitudes, whether in liquid or solid form. Below the altitude of glaciers all high mountains of the Earth have been formed by flowing water, derived from the melting of snow and glaciers, from the runoff of precipitation at the surface, or from fountains. For humans, water in high mountains has various functions. Streams, lakes, and cascading waterfalls became central to the romantic image of the high mountains in the eighteenth and nineteenth centuries, which even persists until today. For others, the water of high mountains is an inexhaustible source of energy, a drinking water reservoir for millions, or an ideal environment for kayaking and river rafting. Large differences in height occurring in conjunction with deeper situated layers of unconsolidated sediment from debris or moraines also present a dangerous torrent hazard which has been threatening the lives and belongings of people in the mountains since immemorial time. To rectify this, as well as the flooding of major rivers in valleys, the world's civil engineers have taken measures. Therefore, the shaping of the high mountains by water is in many regions often followed directly by landscape shaping of humans.

10.1 Torrents

A torrent refers to a permanent or temporary flow of water down a steep slope with a rapid swelling, but not a long persistent episode of flooding. Dangerous amounts of loose rocks or bed load can be moved and deposited either inside or outside the river bed or brought to another stream. Therefore, during a thunderstorm, a more innocuous-looking mountain torrent can suddenly be transformed into a raging river with debris flows.

Central to their dangerousness is the debris load carried by torrents, which together with broken trees and shrubs can be the starting material for mudflows. The debris load may result from weathering of rocks, so that usually the input of solid particles takes place in constant doses and no sudden input comes about. If however the solid input is sudden and takes place in very large quantities, debris flows occur and often far-reaching devastation can result. Such sudden input of debris masses occurs during prolonged rainfall or heavy rain storms. The origin of the input material can be scree slopes, moraines, and other glacial and glaciofluvial deposits. They feed the torrents often by mass movements in the form of landslides.

The world's existing classification scheme for streams (it would take too long to discuss it further at this point) was established mainly within the context of geological and morphological conditions of their basin surroundings. In the alpine region for example, a distinction is made between streams in valley infillings, streams in residual debris, or torrents in hard sedimentary rocks. Furthermore, among the criteria is the possibility of human intervention. Only the portion of precipitation, which is not retained by soil and vegetation, can contribute to torrential disasters. Spruce forests evaporate twice as much water as unwooded areas and can therefore hold back more water in the event of heavy rainfall. The preservation of mountain forests is therefore a priority for the efforts to protect mountains from flooding, landslides, and mudflows.

In addition, other factors can reduce the water retention of vegetation and soil and thus increase the floodwater discharge and debris flow hazard. These include soil compaction due to excessive grazing, heavy agricultural machinery, or the construction of ski runs. Alpine skiing involves the operation of heavy tracked vehicles like snow cats grooming the slopes or pushing by caterpillars for the construction and expansion of slopes. Among the consequences are far-reaching damages to vegetation and soil. In case of insufficient thicknesses of snow cover, the use of snow groomers leads to extreme soil compaction and destruction of the natural soil structure. Therefore, more precipitation flows off the surface. Besides greater erosion, more water enters the streams during heavy precipitations and thus increases the removal efficiency of running water and consequently its sediment load. Any interference in vegetation cover leads to exposed debris and as a result erosion through water, landslides, and other mass displacements occurs.

If the rainfall amount, which increases erosion and landslides, is not impaired in the catchment area of streams by silviculture or bioengineering, it is important to limit the negative impact of the torrent work for cultivated land. This is partly possible due to a drainage targeted to vulnerable areas. There, however, where steep stream sections lead to the fact that the white water attacks over its entire bed width at the stream bottom and slopes, river training structures can be used. By barriers the white water is stripped of this graduation energy and the too steep gradient is replaced by a safe sedimentation gradient; at other places consecutive flood-control retention basins were built (Figs. 10.1, 10.2, 10.3, 10.4, and 10.5).

10.2 Mountain Torrents

Mountain torrents are an essential element of landscape forming processes in the high mountains. The accumulative erosive effect of the river water modifies both the river bed itself and the surrounding landscape of high mountains. Streams and rivers arise by rainfall and by the melting of snow and glaciers. There exist constantly flowing, periodic, and occasionally flowing water streams and rivers. They are known as perennial (from the Latin perennis = constant), intermittent (from Latin intermittere = lying in between),

Fig. 10.1 Mountain torrents. Mangfallgebirge, Bavarian Alps, Germany. For many people who travel to the mountains each year, the raging waters of mountain streams are to the wildly romantic picture of high mountains same as the glaciers. Therefore, we find photographs of mountain streams in travel brochures and information sheets of the tourism industry. For the people living in high mountain regions, this is rather a deceptive picture. Because it is not rare that the rushing mountain streams become a dangerous debris flow assisting torrent. Published with kind permission of © Dr. Alexander Stahr 2014. All Rights Reserved

Fig. 10.2 Moraines and gravels of the ice age. Chiemgau, Germany. Moraines and gravels of the ice age glaciers are frequently the source material for the sediment load of torrents. During heavy rainfall the material reaches floodwater bearing rivers by rinsing, landslides, or other types of mass movements. The increased supply of material is often a product of human impacts on vegetation, particularly in forests since the material is directly exposed to erosion. Published with kind permission of © Dr. Ewald Langenscheidt 2014. All Rights Reserved

and episodic rivers (from the Greek epeisódion = newcomer), respectively.

The water supply of mountain streams is determined by climatic conditions under which special weather conditions or weather events can happen. There is extreme heavy rainfall or prolonged drought, but also the course of snow and glacier melting can lead to flood or drought periods with low water or totally dry riverbeds. In addition there are factors that are important for the retention of rainfall. These include the nature of the subsurface with its rocks

Fig. 10.3 Bank slope demolition. Northern Limestone Alps, Germany. The erosion by mountain streams and torrents on a meander-undercut slope leads to abrasion of loose material and to undercutting of the slope bank. When the river floods and debris flow formation in this way are fed by additional material, it increases its solid content and its hazardousness. Landslides or rinsing fluid movements can also contribute to the expansion of bank demolition. Published with kind permission of © Dr. Alexander Stahr 2014. All Rights Reserved

Fig. 10.4 Slope debris as the origin of river sediments. Berchtesgaden Alps, Germany. In addition to purely glacial and fluvioglacial deposits such as moraines or gravel, material from other debris fall processes is the origin of river bed or torrent loads. Published with kind permission of © Dr. Alexander Stahr 2014. All Rights Reserved

and soils, as well as the type of vegetation cover, the intensity of the water saturation of soil, and the occurrence of freezing conditions before a precipitation event. In the absence of negative human influences on vegetation and soil in high mountains of the middle latitudes, precipitation can be absorbed faster by the ground and stored longer than in desert mountains. For after severe drought, the soils of arid areas are largely filled with air and coated with dust. Their resistance to water prevents rapid infiltration by rainfall and often causes strong surface runoff culminating in catastrophic flood waves. Despite being concentrated in such dry mountains as the more than 3,400 m high North African Tibesti, precipitation and runoff can as sporadic events lead to formation of a distinct canyon relief. Furthermore, in the subtropical highlands, the floods which occur during the first monsoons are entirely due to dry soils resulting from prolonged dry periods.

Snow and glacial melting are of outstanding importance for the flow regimes of countless mountain rivers. Snow stores the seasonal rainfall within glaciers over decades, centuries, and sometimes thousands of years. Therefore, glacial meltwater is of central importance for humans as a source of energy, drinking water, or water for the irrigation of agricultural crops.

A typical glacial stream discharges more or less water depending on the season. Also during the summer months

Fig. 10.5 River training structures. Dolomites, Italy. To limit the negative consequences of torrents on cultivated land, river training structures are used worldwide by the relevant technical authorities, in addition to silvicultural and bioengineering measures. Through the installation of many consecutive barriers white water is robbed of its energy, and a steep gradient is replaced by a safe gradient of sedimentation. Published with kind permission of © Dr. Alexander Stahr 2014. All Rights Reserved

there are striking differences in water supply at different times of day. In the morning the glacial streams flow relatively quietly, as they fall on the evening of a hot summer's rich radiation thundering into the valley. With increasing air temperature and incident solar radiation the amount of water increases over the course of day and reaches its maximum in the late afternoon to early evening. In cooler weather, or on a day in late fall the streams flow quietly again. These characteristic variations in water flow over the course of day and year are determined mainly by melting processes of the glaciers. In the Alps, the largest water flow from the glacial streams occurs in the summer months of June, July, and August. The mountain streams that do not originate from glaciers have their runoff peak in May and June during the snow melt.

Usually the fluvial activity (from the Latin fluvius = river) and the formation of landscape by rivers start at the periglacial altitude of the mountains. This is particularly evident in many mountain rivers, which are caused by glaciers. Even on the glaciers themselves, meltwater collects itself below the firn line. It falls through gaps or holes to the base of the glacier, where it begins its erosive work and forms meltwater channels, potholes, or deep sub-glacial gorges gnawing into the rock. The stream flowing out of the glacier transports moraine material, weathering debris, and fine rock material (glacier milk) to the valley and forms the surface of its bed by depositing these eroded rock fragments. Just before the end of the ice ages coarse gravel and sand are deposited from the glacial streams, forming a so-called gravel floor.

During glacial times streams flow into the valleys and other streams come from tributary valleys. Gradually they become mountain rivers and their transportive and erosive forces are constantly increasing by more confluence out of their catchment area. The entrained debris, often with an audible clatter, scrubs from the bedrock which is polished in places. Larger components of the sediment load are moved along the bottom by flotation. This means that they are dragged along the ground and rolled along the bed. If debris is hopping and jumping in the riverbed, one speaks about moving by saltation (from Latin saltare = to dance). Here, the riverbed is scraped and very fine grains are caused by the mutual attrition of stones; in addition to glacier milk they are carried as suspended load. The ratio of the debris transport sometimes changes greatly. Fast flowing mountain streams carry relatively large amounts of gravel, while the large rivers of the main mountain valleys or the foreland are commonly dominated by very fine material.

The presence of obstacles in the riverbed causes whirls or water vortices to develop, in which entrained debris grinds potholes in the rock. Where the fluvial activity in high mountains is not yet in a position to create a uniform inclination, rapids or waterfalls result at the numerous convex kink points in the river bed or at higher steps. Waterfalls or cascades are also often situated between a glacial hanging valley and the adjacent main valley. In many alpine valleys in winter we see frozen waterfalls that look like giant icicles on the steep valley walls. They are a paradise for extreme ice climbers.

As a result of fluvial erosion (and with the support of glacial erosion) certain alpine valleys were dug out during the course of about million years and reach depths of up to 1,000 m. The shaping of the valleys is caused not only by erosion by streams or glaciers, but also by the planar ablation or denudation and thereby by forming the valley slopes backwards.

Due to the high relief, the transportation force of high mountain streams is very high. If the transportation force decreases rapidly, a large accumulation of loose material takes place. This is particularly the case in glacial U-shaped valleys, which in the Alps are often filled up to a depth of several 100 m with late glacial and river sediments including gravels, sands, and clays. The resulting depositional surface often hides the former U-shaped profile. Other more obvious products of fluvial accumulation, which form a triangular layout as alluvial fans, are also situated where the transportation force of a river decreases suddenly. This is usually achieved by a decrease in surface inclination at the outlet of a tributary river into a large valley or onto a plane. The filling of an alluvial fan is similar to adepositional infilling, but with one essential difference. Valley infillings are bordered by the valley walls or slopes. In contrast, alluvial fans have no such lateral boundaries, so the river's sediment load pours from its exit point out into the foothills in all directions. Hence, these accumulations develop into a fan shape. Some alluvial fans, such as those in the Upper Rhone Valley, are so large that they host settlements. As the river that has constructed an alluvial fan needs to bypass its deposited gravel, it often splits into many smaller arms. They migrate over time across the surface of the alluvial fan, driven especially by seasonal changes in the water discharge. Frequently alluvial fans are the starting points for mudslides.

Related to the alluvial fans are deltas. The difference between the two landforms is that in the case of an alluvial fan the accumulation takes place on a plane or at the mouth of the tributary to a main valley, while at a delta the sediment is carried out into the sea or a lake. In case of heavy debris production and debris infilling in the main valleys a depositional bottom is formed. In contrast to typical V-shaped valleys where the flanks are directly adjacent to the river bed, the bottom extends between the valley slopes and the river bed. There exists a more or less broad valley bottom with the flood plain immediately on either side of the water. At high waters the river banks consisting of gravel and sand deposits were set back, making the riverbed wider and the river more shallow. The consequence is a reduced transportive force of the flowing water. The eroded material from the banks is now deposited as sand or gravel banks once again, only after a short distance. This makes them barriers to the transportation of the river's load, so they absorb more material. An increase in their surface area is the direct result. Finally, they become so high that the stream flows around them in several arms. This process is called degeneration of the stream; it is now named a braided stream. The valley bottom infill may reach such proportions by gravel spill that smaller streams and tributaries are completely blocked by the debris and sometimes go underground. Only at flood water they can flow and transport debris, rubble stones, and finer material.

Through the formation of meanders (Greek Maiandros = Menderes River in Asia Minor, which flows into the Aegean Sea) streams and rivers flow in large loops through the meadows of a valley floor, eroding downwards through its rocks and sediments. Whether a river meanders or flows wild depends on the vulnerability of its banks to erosive processes. Shores consisting of gravels and sands with silt and clay tend rather to form a wild river bed.

In the valley slopes of many high mountain regions, flat surfaces of variable size and shape attract attention. These may rise to varying heights, from high above the valley floor to just above the river bed and are sometimes tiered or stacked on top of another. In most cases, these are river terraces, which represent the remnants of old valley floors. Their fluvial origin shows that deepening by erosion has been interrupted over time by several phases of embankment erosion or deposition. It can be certainly also the case that artificial terraces from farming or vineyards are located within the valley slope and that many years after their abandonment they become difficult to distinguish from natural river terraces at first glance. In addition, some terrace slope flattening occurs, involving the removal of hard rock that has been dissected out alone because of its resistance against processes of weathering and erosion without the assistance of a stream. Such terraces were mentioned previously as denudation terraces.

The river terraces lying high above the valley floor are mostly basement rock terraces that arose over the course of multiphase tectonic uplift of the young orogenic belts or folded mountains. Erosion downward changed several times from erosion to the flanks, so that frequently in the Alps a succession of differently aged terraces or basement rocks can be encountered high above. These terraces are the result of strong lateral erosion during periods of relative tectonic calm, in which the now absent cover of gravel was removed. After this followed a phase of incision, so that the former basement rocks now appear as a terrace.

Fig. 10.6 Mountain river. Lech, Bavaria, Germany. Mountain rivers are a significant landscape element in many high mountains and they contribute to the shaping of the surface in an important way. For people they provide drinking water reservoirs and a system for irrigation of agricultural crops and as well being essential for the production of power. Published with kind permission of © Dr. Alexander Stahr 2014. All Rights Reserved

Fig. 10.7 Glacial melting water. Briksdalsbre, Norway. Many rivers arise from glaciers in high mountain regions. Published with kind permission of © Herbert Funk 2014. All Rights Reserved

Contrasting with the basement rock terraces are the depositional terraces. They too owe their emergence to the river, which again cuts in a depositional terrace and the former valley floor only remains in the form of terrace relicts of different size at the valley flanks. This process of infill and deep erosion driven by tectonic influences and on climate variability (tectonic and climatic influences can be superimposed) can be repeated several times. Sediment layers caused by strong lateral erosion and subsequent deposition can be interbedded in a complicated manner (Figs. 10.6, 10.7, 10.8, 10.9, 10.10, 10.11, 10.12, 10.13, 10.14, 10.15, and 10.16).

Fig. 10.8 Waterfall. Vöringfoss, Hardangervidda, Norway. In high mountains like the Alps or the Himalayas which are characterized by uplift and an associated reduction of their erosion base, rivers have not been in a position to provide a moderate incline. Published with kind permission of © Herbert Funk 2014. All Rights Reserved

Fig. 10.9 Cascade. Norway. When the water of a mountain torrent falls over several small steps it is called a cascade. Published with kind permission of © Herbert Funk 2014. All Rights Reserved

Fig. 10.10 Gorge. Breitachklamm, Germany. A gorge is characterized by nearly perpendicular valley walls. The river occupies the entire width of the valley. Like most valley shapes that were created by flowing water, they are essentially V-shaped valleys. The course of a gorge often follows geological structures such as fractures or dislocations. Published with kind permission of

10.3 Lakes

Mountains, forests, lakes. These are words that for many people mean vacation, recreation, nature, and wilderness. This is certainly true looking at the countless high mountain lakes in Norway, Canada, or South America. But also in the Alps, which (ignoring that there is no longer pristine wilderness) provide some 4,000 lakes for recreation and sports activities. Think about the well-known Carinthian lakes such as Lake Wörthersee or Lake Garda in the Italian Alps.

The origin of lakes can be attributed to very different processes, as explained in previous chapters. There are lakes in high mountains that take the form of ice dammed lakes, avalanche dammed lakes, kar and tongue ice-scour lakes, lakes dammed by landslides, and artificial barrier lakes or reservoirs maintained for energy and drinking water. There are also lakes that lie between moraines and lakes that fill dead ice formed molds. Most of the types listed above which occur in the Alps are relatively small and flat. The water temperatures are in general quite low, so that biological processes within them take place very slowly. Very large and deep mountain lakes are found among other places in the Andes. Here for example is found the highest known navigable lake, the 3,812 m high Lake Titicaca. Furthermore, on the edge of the more than 2,500 m high Baikal Mountains, there is a lake of same name (Ozero Baykal), which with a depth of 1,620 m is the deepest lake on earth (Figs. 10.17, 10.18, 10.19, and 10.20).

Fig. 10.11 Alluvial fans. The photo shows (*left of center*) the famous alluvial fan of St. Bartholomä, which extends into the Lake Königssee (Bavaria, Germany). It consists mainly of light weathering dolomite debris that is present in the lower part of the east wall of Mount Watzmann. In the (geological) future Lake Königssee will be divided into two by the alluvial fan. At the mouths of rivers that flow into a lake or the ocean, a delta is usually formed in preference to an alluvial fan. In the case of St. Bartholomä, we have a "hybrid form." Above an alluvial fan that extends far below the water level into Lake Königssee and reaches the other side is a delta. However, the accumulation area extends far up towards Mount Watzmann's east wall, so that the characteristic conical shape of the alluvial fan is visible above the delta. Published with kind permission of © Dr. Ewald Langenscheidt 2014. All Rights Reserved

Fig. 10.12 Delta. Lago Maggiore, Ticino, Switzerland. A delta is defined as a mound of fluvial sediments at a river mouth which extends into the sea or a lake. Like an alluvial fan, the shape of a delta is nearly triangular. Different supply conditions and basement rock structures produce different delta shapes. The photo shows the delta at the mouth of the river Maggia, which today channels into Lake Lago Maggiore. Published with kind permission of © Dr. Ewald Langenscheidt 2014. All Rights Reserved

Fig. 10.13 Valley bottoms. Lech, Austria. In case of heavy debris production and debris infilling in the main valleys, valley bottoms were formed by streams. In contrast to the characteristic V-shaped valleys, where the flanks are immediately adjacent to the river bed, the bottom extend between the slopes and the riverbed, a more or less broad valley bottom. Published with kind permission of © Dr. Alexander Stahr 2014. All Rights Reserved

Fig. 10.14 Braided streams. Maruia River, New Zealand. At high water the river modifies the embankments of loose gravel and sand deposits. As a result the riverbed becomes wider and shallower. This causes a reduction in the transportive force of the water. The loose worn-out material is deposited again after a short distance as sand or gravel banks. Thus, these materials become obstacles within the riverbed and start collecting further sand and gravel themselves. An increase in their surface area is the result. Finally, the sand or gravel banks become so high that the river must flow around them in several arms. We speak also therefore of wild rivers. This process is repeated again and again at high water, so that the appearance of a braided stream is always taking new forms. Published with kind permission of © Dr. Ewald Langenscheidt 2014. All Rights Reserved

Fig. 10.15 Meanders. Due to the formation of meanders, streams and rivers flow in wide loops through the meadows of valley floors. Whether a river meanders or becomes braided depends largely on the stability of its banks. Meanders can arise by the deposition of gravel to form sand bars, which can further decrease the river's flow velocity. These accumulations can make the river migrate sideways, as other obstacles such as the mouth of a tributary or a large boulder can do. But there are also some hydraulic reasons for meandering. As the river lengthens its bed, its flow velocity and gradient will be reduced. Meandering therefore represents an approach towards an optimum state, where the river discharges the maximum possible volume of water at the minimum possible velocity. Irrespective of the cause of meanders, the loops expand gradually, as can be seen at the *lower right of the photo*. Here, continuous erosion takes place and sometimes undercut slopes can arise. At the inner bank of such a river loop, the flow velocity is much lower, often against the main stream. Sediments are deposited and they form an approximately flat slip-off slope. A breakthrough of such a loop by the water forms an oxbow lake. Published with kind permission of © Dr. Alexander Stahr 2014. All Rights Reserved

Fig. 10.16 River terrace. Ramsau Valley, Berchtesgaden Alps, Germany. In many high mountain valleys, flat floors of varying size and shape can be recognized protruding from the valley walls, which rise just slightly above the river bed, sometimes looking like stairs. As a general rule, these are river terraces, which represent the remnants of old valley floors. Their emplacement at different levels by the erosive action of rivers shows that deepening of the valley floor was interrupted by several phases of deposition. Published with kind permission of © Dr. Ewald Langenscheidt 2014. All Rights Reserved

Fig. 10.17 Dead ice lakes. Lake Hopfensee, Bavaria, Germany. At the end of the ice age many glaciers lost contact between parts of their tongues and the rest of the ice. Gradually, the remaining ice was covered by sediments. After it finally melted a mold was left, which is now filled with water. Published with kind permission of © Dr. Alexander Stahr 2014. All Rights Reserved

Fig. 10.18 Ice-scour lake. Lake Königssee, Berchtesgaden Alps, Germany. The glacier and its tributaries have deepened the former valley, which has followed a fault zone. Published with kind permission of © Dr. Alexander Stahr 2014. All Rights Reserved

Fig. 10.19 Lake dammed by rockfall. Lake Hintersee, Berchtesgaden Alps, Germany. The lake was dammed by rockfall material during the Bronze Age about 3,500 years ago. Published with kind permission of © Dr. Alexander Stahr 2014. All Rights Reserved

Fig. 10.20 Reservoir. Sylvensteinspeicher, Bavaria, Germany. Today reservoirs are mostly used for producing power. Published with kind permission of © Dr. Alexander Stahr 2014. All Rights Reserved

How Man Shapes the High Mountains

11

A. Stahr and E. Langenscheidt, *Landforms of High Mountains*,
DOI 10.1007/978-3-642-53715-8_11, © Springer-Verlag Berlin Heidelberg 2015

As shown in previous chapters, there are particular factors responsible for the creation of specific landscape forms. Two main factors have been mentioned so far: the geological background and its structure, behaving as an endogenous factor and the climate which controls the processes of weathering and erosion as exogenous factor. These latter processes operate at the surface and control the runoff regime of rivers and thus the rate of fluvial erosion both downwards and sideways.

The mountains of the Earth would never have arisen without the tremendous forces of tectonic movements and volcanic eruptions. High up on the ridges only climatic influences such as snow and wind shape the mountains, for example, for mountaineers the so dangerous cornices. Trough valleys would not exist without glaciers, and the water of the Colorado River has over millions of years created the mighty Grand Canyon. These are just a few examples which illustrate the principle of cause and effect as it pertains to geomorphology, the science that deals with the formation and evolution of relief.

In some of the previous chapters it was also made obvious that the human influence on landscapes forming in high mountain regions is significant. Despite emerging relatively late, it is generally recognized that man has to be regarded as a major (and important increasingly) contributing factor to the evolution of relief. The arm of earth sciences which deals with the influence of man on the earth's relief (and which is therefore ultimately concerned with artificial surface forms) is called anthropogenetic geomorphology (from Greek anthropos = human and genés = coming). By artificial forms we mean those that are created directly by man and which involve no natural geomorphological processes in their creation. Artificial landforms serve the purposes of settlements, industries, transports, and the human need for protection against natural (or in the case of military) human forces. The mountain war of 1915–1917 left countless tunnels, caverns, and trenches in the Dolomites and south-eastern Alps. On April 17, 1916, the summit of a mountain, today the 2,462 m high Col di Lana in the eastern Dolomites, was completely blown away with 5,024 kg dynamite and thus reshaped completely in the short space of few seconds. In the Berchtesgaden Alps (Germany) were found numerous bomb craters from the Second World War, looking confusingly similar to natural karst forms. Among artificial landforms (in the truest sense of the word) we must also include the "waste mountains" that can be seen in some high mountain valleys as landfills. Finally, we must also count the various prehistoric tombs and religious places of worship to the shortlist of man-made landscapes.

On all continents humans act as landscape formers. This is especially true in the high mountains, but with great difficulty and hardship. To exist in this initially hostile and

Fig. 11.1 Quarries. Tyrol, nearby Kufstein, Austria. Mining (open pit), which in its broadest sense includes the excavation of material by human activity in quarries, has formed artificial incisions in the earth's surface. The work done these days with the most modern machines leaves behind literally gnawed mountains. In quarries, raw materials are mined for ore, for construction industry, and sometimes also for chemical industry. These raw materials include (among others) rocks such as sandstones, granites, marbles, or limestones, which in addition to its use as building material are used for cement production and also as fertilizer. Published with kind permission of

Fig. 11.2 Roads. Grossglockner High Alpine Road, Austria. The system of roads, which as winding roads override even the steepest natural barriers in high mountains, requires substantial investment due to soil and rock movements. Thus, these roads are referred to as artificial landscapes, considering the relief-changing nature of their construction alone in addition to the presence of the roads themselves. Published with kind permission of © Dr. Alexander Stahr 2014. All Rights Reserved

Fig. 11.3 Tunnels. Tunnels are underground man-made structures and were built for transportation (road, rail, etc.) through mountain ranges, as well as under rivers, straits, or urban settlements. Tunnels — of different nature — are known from the very ancient times (Mesopotamia, Babylonia, Greece, Roman Empire), and they were excavated as part of the construction of ancient water supply systems. The construction of tunnels in mountains is closely associated with roads and railways — think of closed high alpine passes during winter. No other range of high mountains on earth is so full of tunnels as the Alps. There is more than 720 km of tunnel length. There is also, according to CIPRA (Commission Internationale pour la Protection des Alpes), 106,000 km of main and side roads. To illustrate the influence of man on the high mountains of the Alps, here a few figures are given: 300,000 km of service roads, 8,000 km of tracks, and 300 airfields. The photo shows the 311 m long Hochtortunnel at the Großglockner High Alpine Road, Austria, situated 2,505 m above sea level. In the middle of the tunnel is the border between the provinces of Salzburg and Carinthia. Published with kind permission of © Dr. Alexander Stahr 2014. All Rights Reserved

Fig. 11.4 Penstock. Saalfelden, Austria. As part of the energy production system which utilizes the water of streams and rivers, not only dams but also underground and surface pipelines have been built. From a purpose-built water reservoir (and via high pressure pipelines) the water runs downhill through the turbines of a power station. The landscape is as easy to see marked by these artificial forms. Published with kind permission of

Fig. 11.5 Buildings. Neuschwanstein, Bavaria, Germany. People from prehistoric times made their homes in cavities, in caves, and in cave dwellings within rocks which could be carved quite easily. Over thousands of years, this form of settlement has changed fundamentally. Probably only the very least would voluntarily live in a residential pit. Today, the majority of man lives in urban or rural settlements, which we must call anthropogenic landscapes. Through history, only a few had the luck to live in castles, which frequently have altered the landscape themselves, often on a large scale. Published with kind permission of

Fig. 11.6 Terrace cultures. Kali Gandaki Valley, Nepal. For centuries Nepalese farmers have terraced the steep slopes and thereby given them a distinctive and unmistakable shape. The photo illustrates the principle: originally existing arable land is increased in size by elaborate terracing and soil erosion is nearly stopped. The terraces, however, require constant maintenance. Published with kind permission of © Dr. Johannes Thomas Weidinger 2014. All Rights Reserved

Fig. 11.7 Tourism. Schwägalp, Switzerland. Cable cars are part of the tourist infrastructure in the mountains. There are about 12,000 lifts in the Alps. Often forestry operations, such as clearing of forest aisles, are needed to create the conditions necessary for the construction of a cable car as the photo clearly shows. The suspension cables hang from huge masts and these masts must therefore be built upon solid foundations. This is an example of human intervention in the natural condition of soils. Published with kind permission of © Dr. Alexander Stahr 2014. All Rights Reserved

Fig. 11.8 Millstone quarry. Hinterhör, Upper Bavaria, Germany. The medium to coarse-grained sandstones of the millstone quarry were deposited at the turn of Cretaceous — Paleogene period. Quartz and feldspar have been washed (as weathering products) from mainland areas or islands into the surrounding sea. In the quarry here, from the sixteenth to the nineteenth century, millstones were cut out. Holes and cracks were created and then the millstones blasted out by swelling wood. The impressive footsteps of the millstone mining efforts are well preserved. Published with kind permission of © Dr. Ewald Langenscheidt 2014. All Rights Reserved

menacing area at all, man has had to intervene within the existing ecosystems and transform them to his needs. Through the introduction of various agricultural procedures (draining and land reclamation of large valley floors, clearing of forests in several altitude zones for firewood, and creation of pastures by increasing alpine meadows through clearing) man has changed and transformed natural landscape almost completely, for example, a reservoir located high above human settlements, a railroad embankment running through the middle of the mountain regions, or a highway bridge which connects the two sides of a valley, to name a few. These are all artificial features created by man. We encounter them in the mountains in everyday life while hiking or climbing. The people of the high mountains, like those inhabiting other regions of the world, have formed their immediate environment into a cultural landscape that has changed constantly over millennias.

The motivation for man's opening up of mountain regions (assisted by the deployment of advanced technology) is primarily economic. He has been, on the one hand, anxious to exploit the existing agricultural facilities, energy sources, and natural resources. On the other hand, well-developed and secure trade routes across the mountains have always been required. And last but not least, the mountains have become a major recreation area, available to people from urban areas, where they can find rest or confront the ferocity of nature themselves — if they still can be found. In establishing infrastructure amongst well-developed high mountains like the Alps or the volcanic islands of the Hawaiian archipelago with more than 4,000 m high peaks of solidified lava, man has intervened in a relatively short time. However, he has done so with such vehemence that in many places the ecological balance and cultural identity are threatened by mass tourism, urbanization, traffic, questionable economic pursuits, and naturalization of foreign animal and plant species or by increased erosion (Figs. 11.1, 11.2, 11.3, 11.4, 11.5, 11.6, 11.7, and 11.8).

Well, that was it. Now it is up to you to go out and see what you will find ... And you will understand a little bit more what happened outside, in the last 10,000 years, during the ice ages, or during the times before. Have a nice hike and good luck!

Picture Credits

Dr. Alexander Stahr: Chapter opening figures in Chaps. 1, 2, 4, 6, 7, 8, and 11.

Dr. Ewald Langenscheidt: Chapter opening figures in Chaps. 3, 5, 9, and 10.

A. Stahr and E. Langenscheidt, *Landforms of High Mountains*,
DOI 10.1007/978-3-642-53715-8,

About the Authors

Dr. phil. nat. diploma Geographer Alexander Stahr, born in 1959, studied geography, geology, soil science, and botany in Frankfurt am Main (Germany) at the Johann Wolfgang Goethe-University. He is engaged in high mountains since 1986. Dr. Stahr wrote several books as well as numerous professional articles about the Alps and other high mountains. In 1996 he was conferred a doctorate of geosciences by the Johann Wolfgang Goethe-Universität with a soil science-geomorphological thesis on the Berchtesgaden Alps.

Dr. rer. nat. diploma Geologist Ewald Langenscheidt, born in 1955, studied geology, paleontology, and mineralogy at the Philipps-University in Marburg an der Lahn (Germany). He is engaged in high mountains since the end of the 1970s, was an employee in the UNESCO project MAB 6 "Impact of Human Activities on Mountain Ecosystems," and coordinated the multipurpose project "Bodenforschung im Deutschen Alpenraum" (Soil Research in the German Alps) of the German Federal Ministry of Research and Technology. Dr. Langenscheidt accomplished the geological map of the Berchtesgaden National Park and published several books as well as scientific contributions with different aspects to the geology of the Northern Limestone Alps.

A. Stahr and E. Langenscheidt, *Landforms of High Mountains*,
DOI 10.1007/978-3-642-53715-8, © Springer-Verlag Berlin Heidelberg 2015

Further Reading

Abbott PL (2011) Natural disasters. McGraw-Hill Science/Engineering/Math, New York

Abele G (1974) Bergstürze in den Alpen - ihre Verbreitung. Morphologie und Folgeerscheinungen. Wissenschaftliche Alpenvereinshefte 25

Ahnert F (1996) Einführung in die Geomorphologie. Eugen Ulmer, Stuttgart

Amman W, Buser O, Vollenwyder U (1997) Lawinen. Birkhäuser, Basel

Anderson RS, Anderson SP (2010) Geomorphology: the mechanics and chemistry of landscapes. Cambridge University Press, Cambridge

Bätzing W (2003) Die Alpen – Geschichte und Zukunft einer europäischen Kulturlandschaft. Verlag C.H. Beck, München

Bätzing W (2005) Bildatlas Alpen: Eine Kulturlandschaft im Porträt. Wissenschaftliche Buchgesellschaft, Darmstadt

Bennett MM, Glasser NF (2009) Glacial geology: ice sheets and landforms. Wiley, New York

Bishop M, Shroder JF (2004) Geographic information science and mountain geomorphology. Springer Praxis Books, Berlin

Boggs S Jr (2009) Petrology of sedimentary rocks. Cambridge University Press, Cambridge

Bögli A (2011) Karst hydrology & physical speleology. Springer, Berlin

Bosellini A (1996) Geologica delle Dolomiti. Athesia, Bozen

Bull WB (2007) Tectonic geomorphology of mountains: a new approach to paleoseismology. Blackwell, New York

Bull WB (2011) Tectonically active landscapes. Wiley, New York

Bunza G (1982) Systematik und Analyse alpiner Massenbewegungen. In: Schriftenreihe des Bayerischen Landesamtes für Wasserwirtschaft, vol 17. Bavarian State Office for Environment, Munich, pp 1–84

Catt JA (1992) Angewandte Quartärgeologie. Enke, Stuttgart

Erismann TH, Abele G (2001) Dynamics of rockslides and rockfalls. Springer, Berlin

Fossen H (2010) Structural geology. Cambridge University Press, Cambridge

Haeberli W (2007) Changing views of changing glaciers. In: Orlove B, Wiegandt E, Luckman B (eds) The darkening peaks: glacial retreat in scientific and social context. University of California Press, Berkeley, CA

Haeberli W, Kääb A, Hoelzle M, Bösch H, Funk M, Vonder Mühll D, Keller F (1999) Eisschwund und Naturkatastrophen im Hochgebirge. Zürich

Haeberli W, Hoelzle M, Paul F, Zemp M (2007) Integrated monitoring of mountain glaciers as key indicators of global climate change: the European Alps. Ann Glaciol 46:150–160

Haeberli W, Keller F, Krüsi B, Egli M, Rothenbühler C, Meilwes J, Gruber S (2007) Raum-zeitliche Informationen über schnelle Klimaänderungen in hochalpinen Umweltsystemen als strategisches Werkzeug für Analyse, Kommunikation, partizipative Planung und Management im Tourismusgebiet Oberengadin. Schlussbericht des Teilprojekts GISALP, NFP 48 (Nationales Forschungsprogramm Landschaften und Lebensräume der Alpen)

Hantke R (1992) Eiszeitalter - Die jüngste Erdgeschichte der Alpen und ihrer Nachbargebiete. ecomed Fachverlag, Landsberg/Lech

Harvey A (2012) Introducing geomorphology: a guide to landforms and processes. Dunedin Academic Press, Edinburgh (Kindle Edition)

Hsü KJ (1994) The geology of Switzerland: an introduction to tectonic facies. Princeton University Press, Princeton, NJ

Huggett RJ (2007) Fundamentals of geomorphology. In: Routledge fundamentals of physical geography. Routledge, London

Johnson MRW, Harley SL (2012) Orogenesis. Cambridge University Press, Cambridge

Krainer K (1994) Die Geologie der Hohen Tauern. Wissenschaftliche Schriften. Nationalpark Hohe Tauern, Carinthia, Klagenfurt

Kuhle M (1991) Glazialgeomorphologie. Wissenschaftliche Buchgesellschaft, Darmstadt

Kuhle M (2013) The uplift of high Asia above the snowline and its glaciation as an albedo-dependent cause of the quaternary ice ages. Nova Science, Hauppauge, NY

Langenscheidt E (1994) Geologie der Berchtesgadener Berge. Eine Einführung in Stratigraphie, Fazies und Tektonik. Berchtesgadener Anzeiger

Langenscheidt E (2002) Geologische Forschung im Nationalpark Berchtesgaden. In: Nationalpark Berchtesgaden [Hrsg]: Forschungsbericht 46. Berchtesgaden, 57–66

Langenscheidt E, Stahr A (2011) Berchtesgadener Land und Chiemgau: Eine Geschichte von Bergen, Tälern und Seen. Spektrum Akademischer Verlag, Heidelberg

Mangelsdorf J (1990) River morphology: a guide for geoscientists and engineers. Springer series in physical environment. Springer, Berlin

Messerli B, Ives JD (1997) Mountains of the world. CRC Press, Boca Raton, FL

Möbus G (1997) Geologie der Alpen. Sven von Loga, Köln

Owens P, Slaymaker O (2004) Mountain geomorphology. Taylor & Francis, New York

Pettijohn FJ, Siever R (2008) Sand and sandstone. Springer, Berlin

Ritter DF, Kochel RC, Miller JR (2011) Process geomorphology. Waveland, Long Grove, IL

Schönenberg R, Neugebauer J (1997) Einführung in die Geologie Europas. Rombach, Freiburg im Breisgau

A. Stahr and E. Langenscheidt, *Landforms of High Mountains*,
DOI 10.1007/978-3-642-53715-8, © Springer-Verlag Berlin Heidelberg 2015

Semmel A (1984) Geomorphologie der Bundesrepublik Deutschland. Geographische Zeitschrift, Beihefte, Franz Steiner, Stuttgart Wiesbaden
Stahr A (1997) Bodenkundliche Aspekte der Blaikenbildung auf Almen. Nationalpark Berchtesgaden, Forschungsbericht 39, Berchtesgaden
Stahr A, Hartmann T (1999) Landschaftsformen und Landschaftselemente im Hochgebirge. Springer, Heidelberg
Stanley SM (2008) Earth system history. W. H. Freeman, San Francisco
Summerfield MA (1991) Global geomorphology. Pearson, Harlow

Nowadays there are numerous specific articles in scientific journals as well as books and e-books about geomorphology, geology, glaciers, and glaciation and other aspects we dealt with available in the www. Others look at them in a more general way, so it depends on what you are looking for. Also combined with the areas you are interested, for example, the Alps, the Himalayas, or more locally regions, you will get a lot of results searching in the Web. Therefore, it is very difficult for us to give references in detail and we beg for your appreciation.

MIX
Papier aus verantwortungsvollen Quellen
Paper from responsible sources
FSC® C105338

If you have any concerns about our products,
you can contact us on
ProductSafety@springernature.com

In case Publisher is established outside the EU,
the EU authorized representative is:
Springer Nature Customer Service Center GmbH
Europaplatz 3, 69115 Heidelberg, Germany

Printed by Libri Plureos GmbH
in Hamburg, Germany